RUSI
www.rusi.org

Whitehall Paper 81

The Defence Industrial Triptych

Government as Customer, Sponsor and Regulator

Henrik Heidenkamp, John Louth and Trevor Taylor

www.rusi.org

Routledge
Taylor & Francis Group
LONDON AND NEW YORK

Royal United Services Institute for Defence and Security Studies

The Defence Industrial Triptych: Government as Customer, Sponsor and Regulator
Henrik Heidenkamp, John Louth and Trevor Taylor
First published 2013

Whitehall Papers series

Series Editor: Professor Malcolm Chalmers
Editors: Adrian Johnson and Ashlee Godwin

RUSI is a Registered Charity (No. 210639)
ISBN 13: 978-1-138-14850-5 (hbk)
ISBN 13: 978-1-138-02358-1 (pbk)

Published on behalf of the Royal United Services Institute for Defence and Security Studies
by Routledge
4 Park Square, Milton Park, Abingdon, Oxon OX14 4RN
605 Third Avenue, New York, NY 10017

Routledge is an imprint of the Taylor & Francis Group, an informa business

First issued in hardback 2016

SUBSCRIPTIONS
Please send subscription orders to:

USA/Canada: Taylor & Francis Inc., Journals Department, 325 Chestnut Street, 8th Floor, Philadelphia, PA 19106, USA

UK/Rest of World: Routledge Journals, T&F Customer Services, T&F Informa UK Ltd, Sheepen Place, Colchester, Essex CO3 3LP, UK

Contents

About the Authors

Dr Henrik Heidenkamp is a Research Fellow for RUSI's Defence, Industries and Society Programme (DISP). His current research interests are the role of the private sector in defence, national and international defence-management approaches, and contemporary aspects of European security and defence policy. Henrik is also a lecturer, consultant and frequent media commentator on European security and defence industrial policy (including for *Bloomberg*, the *Wall Street Journal*, *New York Times*, *Gulf News*, *Economic Times*, *Deutsche Welle* and *Reuters*). Further, he acts as a mentor for the RUSI/University of Roehampton Business School PhD Programme.

Dr John Louth is Senior Research Fellow and Co-Director for Defence, Industries and Society at RUSI. He served as an officer in the Royal Air Force for sixteen years before working as a consultant and programme director extensively throughout the defence and energy sectors. His work has included the audit and governance of the UK strategic deterrent and government–industry partnering initiatives. John is also a specialist adviser to the House of Commons Defence Select Committee and a non-executive adviser to NDI Ltd. He teaches at Roehampton University Business School in London, his work is published across a broad spectrum of outlets and he is a regular commentator for the BBC, ITN, Sky News and Al Jazeera networks.

Professor Trevor Taylor is Professorial Fellow at RUSI where he co-directs the Defence, Industries and Society Programme. He teaches for Cranfield University at the Defence Academy, where he was head of department for twelve years, and the Naval Postgraduate School in Monterey, CA. For six years he was an elected council member of the Defence Manufacturers Association and he has also headed the International Security Programme at the Royal Institute of International Affairs.

Acronyms and Abbreviations

AEW	Airborne early warning
AIN	Ausrüstung, Informationstechnik und Nutzung [Equipment, Informational Technology and In-Service Support]
AWACS	Airborne warning and control system
AWG	Außenwirtschaftsgesetz [Foreign Trade Act]
BAAINBw	Bundesamt für Ausrüstung, Informationstechnik und Nutzung der Bundeswehr
BAFA	Bundesamt für Wirtschaft und Ausfuhrkontrolle [Federal Office of Economics and Export Control]
BBP	Better Buying Power initiative
BDSV	Bundesverband der Deutschen Sicherheits- und Verteidigungsindustrie [Federation of German Security and Defence Industries]
BIS	Department for Business, Innovation and Skills
BMVg	Bundesministerium der Verteidigung [Federal Ministry of Defence]
BMWi	Bundesministeriums für Wirtschaft und Technologie [Federal Ministry of Economics and Technology]
BSR	Bundessicherheitsrat [Federal Security Council]
CDE	Centre for Defence Enterprise
CIFIUS	Committee on Foreign Investment in the US
CONDO	Contractor on deployed operations
CONLOG	Contractor logistics
COTS	Commercial off-the-shelf
CPM	Customer Product Management system
CRS	Congressional Research Service
CSCE	Conference on Security and Co-operation in Europe
CSO	Contractor support to operations
DAR	Defense Acquisition Regulation
DARPA	Defense Advanced Research Projects Agency
DASA	Defence Analytical Services Agency
DBS NSV	Defence Business Services National Security Vetting
DE&S	Defence Equipment and Support
DEFCON	Defence Condition
DESO	Defence Export Services Organisation
DIS	Defence Industrial Strategy
DoD	Department of Defense
DRS	Diagnostic-Retrieval Systems, Inc
DSG	Defence Support Group
DSO	Defence and Security Organisation
DSTL	Defence Science and Technology Laboratory
DTS	Defence Technology Strategy

EADS	European Aeronautic Defence and Space Company
EAR	Export Administration Regulations
ECO	Export Control Organisation
EU	European Union
FAR	Federal Acquisition Regulation
FEMA	Federal Emergency Management Agency
FFF	Fähigkeitslücke und Funktionale Forderung [Capability Gap and Functional Requirement]
GAO	Government Accountability Office
g.e.b.b.	Gesellschaft für Entwicklung, Beschaffung und Betrieb [Company for Development, Procurement and Operation]
GOCO	Government owned, contractor operated
GPS	Global Positioning System
HDW	Howaldtswerke-Deutsche Werft
HIL	Heeresinstandsetzungslogistik
IED	Improvised explosive device
IGS E	Industriegruppe Service im Einsatz
IPTs	Integrated project teams
ISAF	International Security Assistance Force
ITAR	International Traffic in Arms Regulation
IUB	Infrastructure, Environmental Protection and Services Directorate
JSF	Joint Strike Fighter
KBR	Kellogg, Brown and Root
KdB	Konzeption der Bundeswehr [Conception of the Bundeswehr]
KMW	Krauss-Maffei Wegmann
KWKG	Kriegswaffenkontrollgesetz [War Weapons Control Act]
LHBw	LH Bundeswehr Bekleidungsgesellschaft
LOGCAP	Logistics Civil Augmentation Program
MBT	Main battle tank
MoD	Ministry of Defence
NAO	National Audit Office
NATO	North Atlantic Treaty Organization
OCS	Operational contract support
OECD	Organisation for Economic Co-operation and Development
OEM	Original equipment manufacturer
OGEL	Open General Export Licence
OIEL	Open Individual Export Licence
OSCE	Organization for Security and Co-operation in Europe
PFI	Private finance initiative
PJHQ	Permanent Joint Headquarters
PMSC	Private military and security company
PPP	Public–private partnership
QDR	Quadrennial Defense Review

R&D	Research and development
SBIR	Small Business Innovation Research programme
SDSR	Strategic Defence and Security Review
SITCL	Single Individual Trade Control Licence
SIEL	Standard Individual Export Licence
SIPRI	Stockholm International Peace Research Institute
SME	Small and medium enterprise
SPD	Social Democratic Party (Germany)
SPOT	Synchronized Predeployment and Operational Tracker
SSA	Special security arrangement
TEPIDOIL	Training, Equipment, Personnel, Infrastructure, Doctrine and Concepts, Organisation, Information and Logistics
TIV	Trend indicator values
TK	Teilkonzeption [Special Conception of the Bundeswehr]
TOBA	Terms of Business Agreement
UAE	United Arab Emirates
UAV	Unmanned aerial vehicle
UCAV	Unmanned combat air vehicle
UK	United Kingdom
UNSC	United Nations Security Council
UOR	Urgent Operational Requirement
US	United States of America
VS	Verschlusssachen
ZV	Target Agreement (Bundeswehr)

FOREWORD

The relationship between the businesses that contribute to the defence and security of the nation state – that is, the defence industry – and government is critical if citizens are to be kept safe and secure. It is, naturally, a relationship that also often underscores a modern state's foreign-policy stance and sense of place in the world. Yet, despite its clear importance, this subject is seldom properly explored or analysed in a rigorous, clear-eyed manner. Not surprisingly, and perhaps as a consequence, government policies relating to defence industry (if they exist at all) can seem somewhat contrived, ill-considered and contradictory.

This Whitehall Paper seeks to capture in a systematic manner the key components and drivers of the relationships that bind a government to its defence industrial base. It is a bold ambition and one that is singularly important as the United Kingdom approaches the next Strategic Defence and Security Review. By looking at the subject through the specific lenses of the United Kingdom, the United States and Germany, the authors have constructed a cat's cradle of insights, lessons and conclusions that will not only help policy-makers and industrialists but, more importantly, inform taxpayers of what they should demand from their politicians and industrialists if democracy's arsenal is to stay healthy and engaged.

In recent years, the Royal United Services Institute for Defence and Security Studies (RUSI) has published a range of significant research papers under the banner of its Defence, Industries and Society programme, not all of them welcomed by government or certain business executives. However, it is clear that the work of understanding industry's role in defence and national security is not just a policy imperative but, in an uncertain and dangerous world, it may also be an existential one. Indeed, the authors report that 45 per cent of the British effort on the front line in Afghanistan was provided by members of the private sector, working shoulder-to-shoulder with their military colleagues. The inevitable conclusion is that without a thriving defence industry, the UK's military commitments and operations (and, indeed, those of many other Western nations) in the twenty-first century do not seem viable.

Dr Heidenkamp, Dr Louth and Professor Taylor have been researching this subject in a nuanced, reflective and balanced manner for a number of years. This monograph demonstrates that extensive learning and, in my view, is a major contribution to our understanding of defence, national security, and the interdependencies and challenges of the defence economy in the new millennium.

The Right Honourable James Arbuthnot MP
Chairman of the House of Commons Defence Select Committee
November 2013

I. INTRODUCTION

Defence industries are still regarded, principally, as a part of a country's arsenal: a repository of goods, services, know-how and manpower from which a military can draw in times of conflict.[1] Indeed, companies such as BAE Systems, Lockheed Martin, Raytheon and QinetiQ supply goods and services for people in uniform to use on operations and will continue to do so for many decades yet. However, companies today are much more than just component parts of a military's supply chain. For example, in the United Kingdom, contractors drawn from the private sector provide 45 per cent of the country's overseas military effort.[2] In Afghanistan, at the height of operations in 2010, sixty-seven companies employing close to 5,000 employees were fielded on the front line in direct support of British operations.[3] Industry is not just democracy's arsenal – it is democracy's operational partner, meaning that much of today's military competencies actually reside in the industrial base. Consequently, understanding the relationship between government and defence industry becomes a critical policy discourse in the provision of national security.

This Whitehall Paper deals with how governments in three countries – the United Kingdom, Germany and the United States – deal with the defence businesses on their territory, across three dimensions: as the customers of such businesses, as their sponsors and as their regulators. These states, which account for three of the largest NATO defence budgets, are used as exemplars so that the lessons and insights gleaned might be applied to other states. The authors have found that, typically, the same element of government – the defence ministry – is the leading voice in all three of these spaces, and a working definition is offered for each.

---

1 See, for example, Trevor Taylor and Keith Hayward, *The UK Defence Industrial Base: Development and Future Policy Options* (London: Brassey's for RUSI, 1989); Jacques S Gansler, *Democracy's Arsenal: Creating a Twenty-First-Century Defense Industry* (Cambridge, MA: MIT Press, 2011).

2 John Louth, 'The Defence Industrial Knowledge Base: The Core Capability?', *RUSI Defence Systems* (Vol. 15, No. 1, August 2012) pp. 42–43.

3 Henrik Heidenkamp, 'Sustaining the UK's Defence Effort: Contractor Support to Operations Market Dynamics', *RUSI Whitehall Report 2-12* (April 2012).

Government as Customer

> **Text Box 1:** Operation *Ellamy*: Industry Earns a Pat on the Back.
>
> 'Twenty UK and overseas defence companies have been congratulated by the Minister for Defence Equipment, Support and Technology, Peter Luff, for the contribution they made to the success of Operation Ellamy – the UK's share of coalition operations in support of the United Nations mandate on Libya.'
>
> Source: The magazine of the UK MoD's Defence Equipment and Support organisation, *Desider* (January 2012), p. 8.

Most obviously, a government buys goods and services from the defence industrial sector and from this perspective, government – and more specifically its armed forces – is the customer of that area of the economy. For firms offering specialised military equipment, such as artillery pieces or warships, the home government may be the only customer for their goods, but in any case is very likely to be the most important customer, as other governments are unlikely to buy a product that the domestic government has not endorsed with a purchase.

One important aspect of all customer–external supplier relations is the shape and predictability of demand. At the macro level, defence budgets in NATO countries are not growing and may even be declining in real terms – a trend that seems unlikely to be reversed in the foreseeable future (the next three to four years). At the same time, demand for specific elements within defence can increase, with current governments seemingly prepared to spend more on cyber-defences (and perhaps even offensive cyber-capabilities), for example.[4] The argument for transparency is, of course, that a customer that signals future intentions helps potential suppliers to shape their investment and research plans in a manner that will benefit their customer.

Governments vary considerably as to how open they are willing to be about their future defence spending plans. The Australian government is particularly open in this regard, publishing Defence Capability Plans which are intended to increase suppliers' confidence about the shape of future demand.[5] The British government is seeking to follow a similar path but, to date, the details of its equivalent Defence Equipment Plan are not available in any useful level of granularity to potential suppliers.[6] Moreover, the Australian capability plans have been unstable and much affected by

---

[4] Robert H Scales, Jr, *Future Warfare* (Carlisle, PA: US Army War College, 1999).

[5] Australian Department of Defence, 'Defence Capability Plan 2012: Public Version', May 2012, <http://www.defence.gov.au/publications/capabilityplan2012.pdf>, accessed 7 October 2013.

[6] UK Ministry of Defence, 'The Defence Equipment Plan 2012', January 2013, <https:www.gov.uk/government/uploads/system/uploads/attachment_data/file/70258/Defence_Equipment_Plan_2012_20130130.pdf>, accessed 7 October 2013.

financial cuts to defence budgets, suggesting they do not necessarily ensure greater predictability. Both Germany and the US champion notions of supplier openness but, as will be shown in subsequent chapters, to very different ends. There is little doubt, however, that the concept of government as 'customer' remains highly relevant to the industry–state relationship.

Government as Sponsor

Text Box 2: The Mutual Benefits of Government Sponsorship.

> 'It has long been recognized that market imperfections are unlikely to provide all the R&D needed in sectors such as defence ... there is a well-established acceptance on both sides of the Atlantic that governments will provide much of the funding, and this makes possible the conduct of international collaborative research programmes mediated by governments, if they see it in their joint national interests.'

Source: Alastair Cameron (ed.), 'Defence Research and Development in the Atlantic Nations', RUSI Occasional Paper, 2007, p. 1.

At the same time as being customers of the defence industrial sector, governments frequently play a sponsorship role, helping firms to survive and prosper. Governments, even in free-market societies, obviously give some direction to the private sector as a whole, not least by investing in infrastructure, making provisions for education, providing state funding for research, giving tax breaks for corporate research-and-development spending, and so on. They also often provide information about foreign markets and support sales campaigns in different areas. The US Department of Commerce keeps staff in many US embassies while the UK Foreign Office recognises the promotion of national economic interests to be a significant part of its mission. In defence, however, there is often a special emphasis on sponsorship of industry through practices of national preference in procurement choices, the funding of research and development, and support for exports. Where the defence industry receives special treatment from government, as arguably is the case with each of the countries examined in this Whitehall Paper, it is because the government sees that sector as one of special significance.

Government as Regulator

Text Box 3: BAE–EADS Merger 'a Missed Opportunity', Says Airbus.

> 'BAE and EADS were forced to abandon £30bn merger talks earlier this month after the German Chancellor Angela Merkel personally intervened to block a deal.'

Source: Angela Monaghan, *Daily Telegraph*, 23 October 2012.

A third, often neglected, element is the directing of defence businesses, which includes national and international regulation. Governments do not just buy from and sponsor defence businesses: they also seek to constrain their behaviour through legislation, regulation and policy. Of particular relevance are policies and processes determining who can be employed in sensitive positions; the control of information (for example, the Official Secrets Act in the UK) and the control of exports (for example, the Export Control Act); and wider corporate behaviour (for example, the Bribery Act). The authors would also suggest that the ownership of corporations, or limitations on the powers of shareholders, is an important area of government regulation.

The Significance of the Debate

At first glance, the three areas of the authors' self-styled defence industrial triptych might appear rather dry, technical fields of interest only to specialists. However, it is argued here that they should be viewed as matters of high politics, of key importance to national and collective military capability.

This monograph is partly about the conflicting pressures in the defence industry between the political – where the sovereign state has been central for over 500 years – and the managerial, with industry's necessary focus on efficiency and effectiveness, economies of scale and access to market. It examines current practices, areas of change, and the dilemmas faced by governments and companies.

The Specialist Defence Sector and its Market Structures

Defining the parameters of the defence sector is not easy, since armed forces need supplies of a wide range of goods and services, many of which (such as diesel fuel, clothing and facilities maintenance) have an essentially civil or dual-use character. The authors' concern here is more limited and is focused on the systems and services that are specifically developed for defence, especially the air, sea and land platforms, surveillance and communications systems, and weapons and munitions that are central to military capability. It is in these areas that much of the money is spent and the majority of problems lie.

The market structures in this domain are rather different to those found elsewhere and are very distant from the perfect market features used to generate much economic theory. In defence, it is often a single government (the customer) that specifies a military requirement, thereby determining what is to be developed and produced. That government normally has to fund much or all of the development costs. This obviously contrasts with most other markets, in which companies considering innovation and product development have to estimate the likely demand

and, whilst market research can help, significant risk remains. Unsolicited innovation does occur in defence but is very much the exception rather than the rule because of the significant timescales, costs and risks involved.

The central trend in the specialist defence sector since the end of the Second World War has been consolidation among suppliers, most obviously within states, so that today even a major state such as the UK or France has only one or two indigenous suppliers in a sector. Even the US Department of Defense, which controls by far the largest defence budget in the world, has only a very limited choice of suppliers as far as ballistic missiles, submarines, tanks, combat aircraft and other military assets are concerned. Given that these firms must sell to their home governments to have much chance of third-party sales, government–industry relations in defence are often about a sole buyer talking to an oligopoly about the costs, delivery dates and performance of a complex product which does not yet exist and involves significant technological and financial risk. Some advanced defence projects, especially in their early stages, often involve 'uncertainty' in terms of hazards that cannot be foreseen or quantified. How this uncertainty is conceptualised and managed between the industrialist and the governmental overseer shades much of the debate around defence acquisition and its reform.

Table 1 below, reporting research done within the UK Ministry of Defence, captures data about the increasing cost of British frigates. The Type 23 design represented a commitment to bringing this upwards trend under control, although the first ships went into service without a fire-control system.

Table 2 offers similar information about British combat aircraft, as well as their increasing speed and range.

Table 3, summarising much of the research, suggests no standard tendency among different sorts of equipment, other than a clear

Table 1: British Frigates, 1956–89.

Class	Earliest In-Service Date	Unit Cost (£ million, 2009 prices)
Type 14	1956	33
Type 12	1957	62
Type 12M	1957	71
Type 81	1959	94
Leander	1961	81
Type 21	1972	192
Type 22	1976	413
Type 23	1989	183

Source: N Davies, A Eager, M Maier and L Penfold, 'Intergenerational Equipment Cost Escalation', Defence Economic Research Paper, 18 December 2012, p. 12.

Table 2: British Combat Aircraft.

Type	In-Service Date	Speed (mph)	Range (miles)	Unit Cost (£ million, 2009 prices)
Hunter	1955	715	1,130	4.29
Lightning	1960	1,049	1,148	4.55
Harrier	1968	843	2,073	8.66
Tornado	1979	854	1,207	29.59
Typhoon	2006	1,550	1,801	66.54

Source: Davies, Eager, Maier and Penfold, 'Intergenerational Equipment Cost Escalation'.

orientation to unit costs rising in real terms, year on year, well above average increases in defence budgets in NATO countries.

Companies that lose a competition for a research-and-development project often decide to leave that sector by selling or closing the business concerned. For example, Hunting Engineering left the defence sector when it lost the contract for the operation of the Atomic Weapons Establishment to Lockheed Martin and Serco.[7]

Another core characteristic of much of the defence industrial sector is the high financial, technological and knowledge-based entry barriers. Consequently, old firms leaving the industry are not usually replaced by new entrants. The American firm General Atomics, which pioneered the development of larger unmanned aerial vehicles (UAVs) and is the manufacturer of the Predator and Reaper systems, is an exception to this rule. In general, the ability to make a modern warship, combat aircraft, smart missiles or even an armoured vehicle depends on such a broad range of factors that companies and countries trying to enter these

Table 3: Simple Unit–Time Trend Model of Overall Intergenerational Cost Growth.

Item	Real Annual Cost Growth (%)	Time Period
Generic naval platforms	2.6	–
Destroyers	4.3	1962–2011
Frigates	2.9	1956–2000
Generic submarines	3.8	–
Aircraft carriers	5.8	1955–2011
Combat aircraft	5.9	1955–2008
Main battle tanks	3.4	1963–94

Source: Davies, Eager, Maier and Penfold, 'Intergenerational Equipment Cost Escalation'.

7 Air Force Research Institute, *Deterrence in the Twenty-First Century* (Maxwell AFB, AL: Air Force Research Institute, 2009). This lays out the proceedings of a RUSI conference held in London on 18–19 May 2009, co-sponsored by the Air Force Research Institute and King's College London.

Table 4: Leading Defence Companies, 2012.

2012 Rank	Company	2012 Defence Revenue	2011 Defence Revenue	Change (%)	Total Revenue 2012	% of Defence Activity	Country
1	Lockheed Martin	44,883	43,978	2.1	47,182	95.1	US
2	Boeing	31,378	30,700	2.2	81,698	38.4	US
3	BAE	26,813	29,130	−8.0	28,255	94.9	UK
4	Raytheon	22,705	23,056	−1.5	24,414	93.0	US
5	General Dynamics	21,023	23,491	−10.5	31,513	66.7	US
6	Northrop Grumman	20,600	21,400	−3.7	25,218	81.7	US
7	EADS	14,913	16,093	−7.3	72,637	20.5	Netherlands
8	Finmeccanica	12,529	14,585	−14.1	22,136	56.6	Italy
9	United Technologies	12,117	11,000	10.2	57,700	21.0	US
10	L-3 Communications	10,839	12,521	−13.4	13,146	82.5	US
11	Thales	9,213	9,493	−2.9	18,256	50.5	France

Source: All revenue given in US$ millions. Adapted from *Defense News* (figures rounded), <http://special.defensenews.com/top-100/charts/rank_2013.php?c=FEAands=T1C>, accessed 7 October 2013.

fields – such as China, India and even Japan – incur enormous expense to make progress that is often much slower than that hoped for. South Korea, for example, has been able to move much more quickly into the global automobile and consumer-electronics sectors than into the defence sector, and its attempted entry into the combat-aircraft world remains late, over-budget and incomplete.

Implications of the End of the Cold War

Industrial consolidation in the defence sector is a saga that has been running for more than seventy years. However, since the end of the Cold War, two further fundamental changes have occurred, with profound and perhaps contradictory implications for government relations with the defence industry.

The Growing Presence of the Defence Multinational

The first of these is that defence consolidation has crossed national borders, creating defence multinationals with development, production and support activities in more than one country. This change has been driven, in part, by the lack of growth in national defence budgets and markets. Indeed, eight of the top eleven defence companies in the world have this characteristic (see Table 4). American defence companies have invested in the UK in order to ease access to the British market. Similarly, many European companies recognise the need to invest in US facilities if they are to be awarded contracts by the Department of Defense (DoD). British companies (BAE Systems, Rolls-Royce, Cobham, QinetiQ and Meggitt) have had the most success in this regard and have been allowed to invest heavily in the US. Other European investors include Finmeccanica (an Italian conglomerate), which succeeded in buying Diagnostic-Retrieval Systems, Inc (better known as DRS) in 2009, and the European Aeronautic Defence and Space Company (EADS) – a continental multinational – which has gained a US presence through success in selling helicopters and its bidding efforts in the US Air Force tanker competition. Thales, as a partly state-owned French company, has experienced greater difficulty in gaining acceptance across the Atlantic because of the perceived influence of the French government over its corporate strategy.

Within the European arena, Thales, Finmeccanica and EADS have invested extensively in the UK. However, whilst the broader defence industrial sector has not been immune to the increasing dominance of the multinational business, France and Germany have resisted foreign involvement in their land and naval defence sectors. In the realm of aerospace, however, EADS and the missile developer MBDA are involved with both countries as key developers of smart weapons and air and space platforms respectively.

Transnational defence mergers occur only with the consent of the government of the state in which external investment is to occur while, to a certain extent, government stances in relation to the defence sphere can affect attitudes to foreign investment in general. The most explicit drivers are in Europe, where, in 1998, heads of government in France, Germany and the UK recognised that Europe's aerospace and defence electronics firms needed to be restructured across borders in order to better compete and collaborate with the large and essentially national businesses of the US.[8] That agreement notwithstanding, and with the important exceptions of EADS and MBDA, it is clear that the UK has been much more open to foreign investment in defence than either Germany or France. The US has encouraged British and, to a lesser extent, other foreign investment in the US defence sector apparently because its authorities want to sustain the scope for competition among businesses operating in the US. However, as is made clear in the chapters on sponsorship and regulation (Chapters III and IV respectively), this has been only on the condition that foreign firms are tightly restricted, especially with regard to exporting technology.

---

8 Henrik Heidenkamp, John Louth and Trevor Taylor, 'The Defence Industrial Ecosystem: Delivering Security in an Uncertain World', *RUSI Whitehall Report 2-11* (June 2011).

The Decline of Nuclear Deterrence and the Changed Strategic Situation
The second area of change since 1990 concerns the high-level military-industrial strategic situation: there has been a greatly reduced emphasis on the (nuclear) deterrence of a known adversary and an increased focus on the capability to undertake surprise and sometimes sustained military operations. Concomitantly, a significant aspect of a state's ability to use its armed forces now lies with assured access to a sufficient and agile supply of defence materiel and capabilities drawn from its own industrial base.

During the Cold War, deterrence was the name of the game and, for Western Europeans especially, this depended on conventional forces being strong enough to prevent an immediate resort to nuclear weapons following a possible Warsaw Pact offensive, but not so strong as to discredit the necessity of using nuclear arsenals after a week or so of fighting. This was the essence of NATO's 'flexible response' doctrine, in which NATO asserted its readiness to initiate the use of nuclear weapons. The West wished to send Moscow the clear message that any resort to aggression could result in the use of nuclear forces. Had NATO countries built their conventional forces to a level where they could aspire to hold Soviet forces indefinitely, this would have incurred massive expense and signalled that the West was extremely reluctant to use nuclear weapons, which in turn would have weakened deterrence. The adoption of this strategy meant that there was no provision for the mobilisation of industry for a Third World War, at least not in Western Europe. This was a conflict that would have been fought relying on equipment, ammunition and spares that were in stock at the outbreak of hostilities.[9]

Since the end of the Cold War, the major governments of NATO have had to discard the concept of deterrence, which offered a known adversary, certain operational parameters and a given playbook of responses to aggression. Now, NATO countries are faced with contingent and uncertain campaigns and operations that may come as a surprise in terms of both their nature and location. To a greater or lesser extent, these operations have required the mobilisation of a supply base and a plethora of Urgent Operational Requirements (UORs) being laid on the private sector. Once again, as before 1945, a country's assured access to capable and agile industrial production has become a significant aspect of its ability to deploy its armed forces. In the UK, this reliance upon industry was recognised implicitly in the British Defence Industrial Policy of 2002 and explicitly in the Defence Industrial Strategy of 2005.

Clearly, this change has been more important for the UK and some other European countries than for the US. During the Cold War, resource

[9] Annually, *The Military Balance* – of the International Institute for Strategic Studies – lists the perceived key variables in this arena: the number of systems in national inventories and the extent of state defence spending.

pressures steadily forced the UK to focus more and more of its defence effort on its NATO responsibilities, with capabilities for 'out-of-area' operations becoming an ever-lower priority in the defence White Papers of the 1980s. The British government did have to mobilise the defence industry in 1982 in order to prepare for and execute the recapture of the Falklands Islands but this did little to increase the government's emphasis on the strategic importance of defence industrial capabilities. The defence-procurement policy theme of that time was to subject British firms to external competition, which implied a readiness to see the weak go to the wall.

The US, on the other hand, was always concerned with a wide range of conflict possibilities and experienced a protracted conventional conflict in Vietnam. Its serious, continuous concern with its defence industrial base was apparent throughout the Cold War and afterwards. In the Cold War debates about deterrence and NATO strategy, the US was most concerned that the Warsaw Pact's advantage in terms of quantity should be largely offset by a Western technological edge, which required extensive spending on research and development. In the event of a failure of deterrence in Europe, it was also in the US's interest that fighting should be limited to the conventional-weapons domain for as long as possible, so as to spare American territory from damage. From a broader, external security-policy perspective, especially after the Nixon Doctrine was first expounded in 1969 – under which the supply of US weapons was to be a substitute for the despatch of American forces to friendly countries and regimes – Washington needed an assured source of development and production of such equipment. For Europeans, too, a final useful aspect of a national defence industrial capability was that it enabled arms exports, either as sales or aid, to serve as a foreign-policy tool.

Clearly, defence industries were also important in Europe during the Cold War because spending defence budgets at home meant employment and tax revenues from the employees and firms concerned, sometimes useful technological gains and savings on foreign exchange. In considering how a domestic defence industry could reduce the economic pain of the defence effort, it should be remembered that in 1990 NATO members spent about twice the share of their Gross Domestic Product on defence that they did by 2013.[10]

With these qualifications, however, the fundamental generalisation remains valid: that during the Cold War the prime defence mission for Western states was deterrence of the Warsaw Pact, which did not involve a need to mobilise industry for combat. After 1990, the new challenge of

[10] NATO, 'Financial and Economic Data Relating to NATO Defence', press release, 13 April 2012, <http://www.nato.int/cps/en/natolive/news_85966.htm?mode=pressrelease>, accessed 7 October 2013.

determining how to undertake surprise and risky missions at relatively short notice meant that assured access to a flexible and responsive defence industrial sector was especially valuable for governments.

Intergovernmental Co-operation and the Persistence of National Feeling

Finally, in the face of rising equipment costs, there has been increasing recognition at the governmental level of the limitations of national defence efforts within Europe, with arguments being put forth in favour of greater pooling and sharing of military resources, national specialisation and more co-operative projects. While the institutional basis for increased European defence co-operation is a contested area – with British governments in particular being wary of the potential roles of the European Commission and the European Defence Agency – the generic pressures towards increased defence co-operation among European states are clear. The UK and France have made a significant move towards greater co-ordination, integration and the sharing of national efforts, with agreements signed in autumn 2010 covering nuclear, as well as conventional, areas.[11]

However, in Europe, the contrast between the demands of popular politics and those of the 'managerial' (or 'functionalist') perspective is stark. Politically, the notion of national identity and independence has a continuing, strong attraction for substantial portions of European electorates. The German refusal to countenance a merger of BAE Systems and EADS typifies this tension. In contrast, the managerial outlook emphasises the value of, indeed the need for, a single European market with a shared set of standards and regulations, and in which the costs and risks of operating a multitude of national currencies are avoided. How else are businesses in Europe to compete with those based in the populous markets of the US, Japan and China, for example?

Thus there is an overall situation comprising the coming together of three uncomfortable bedfellows: the continued emotional and political preference for national identity and capability; the managerial need for defence businesses to be multinational in nature and for governments at least to co-operate on big projects; and the dependence of military capability on a capable and agile military supply.

11 The Anglo-French Lancaster House Treaties were signed in November 2010 by then-President Nicolas Sarkozy and Prime Minister David Cameron. They covered key areas of defence research and development, such as unmanned aerial vehicles (UAVs), satellite communications and submarine technology, devoting an annual budget of €50 million each to joint technological development.

The Purpose of the Whitehall Paper

Looking forward is a hazardous activity, so there is a need to be explicit about the assumptions underlying the discussion put forward in this monograph. Of these, the most important is that nuclear deterrence will not return to its pre-1990 centrality and that the ability to use armed forces will continue to rest on assured access to a supply base.

For the dynamics of the future, much hinges on the key elements of the relationship between governments and defence industries. Accordingly, the role of this volume is to assess government attitudes, policies and behaviour in three NATO states towards the defence industry, and to explore the implications of these for both national and transnational defence firms and for future developments in defence, economic and national-security co-operation. The material here is important at the governmental level for determining prospects for co-operation on specific projects, and for companies it is significant for establishing where to invest in base production and development facilities. Specifically, the chapters that follow were prepared in relation to a series of key questions:

- Do national policies across the three areas of 'customership', sponsorship and regulation enjoy coherence or can contradictory elements be discerned?
- Do government constraints seriously hinder the ability of defence enterprises to operate effectively and efficiently?
- Do governments' stances towards the defence industry facilitate or bring friction to defence collaborative and co-operative projects?
- Do governments' stances towards the defence industry encourage it to invest in some countries rather than others?
- What directions of change can be discerned in governments' stances towards the defence industry?
- Do governments aspire to control (and/or sponsor) the overseas investments of defence companies based on their territory?
- Does government policy and behaviour recognise the implications of and questions raised by cross-border investments in the defence industrial sector?

The authors return to these questions in the concluding chapter, having discussed the role of government as customer, sponsor and regulator of defence industries in the intervening chapters. Moreover, a focus on the US, the UK and Germany is maintained because of the significance of these countries to NATO and the relative ease with which data can be sourced. This is in marked contrast to other Western powers, such as France, and to powers such as China and Russia. However, the authors believe that the lessons gleaned from these countries are relevant to the experiences of other states.

There are no iron laws in this area: in principle, the emergence of more 'traditional' collaborative projects, such as Tornado and Typhoon, could reduce the pressure for the strengthening of transnational defence businesses. On the other hand, the presence of a transnational company, such as EADS (to be known as Airbus Group from 2014) or MBDA, could facilitate agreement on such projects and their subsequent delivery. Increased investment in Europe by American defence businesses could enable them to escape some of the constraints of American export controls if they develop technology in Europe. In contrast, if the US government looks to exercise extraterritorial control over technology developed in Europe but owned by American firms, these could become the agents through which US export-control practices are brought into Europe. The authors can aspire only to throw some light on these questions rather than generate final answers, but intend, at least, to make sure that their significance is recognised.

II. GOVERNMENTS AS CUSTOMERS OF DEFENCE INDUSTRIES

This chapter addresses some of the key dimensions of government's role as the customer. It identifies the functions that government has entrusted to industry in defence, outlines government's perceptions of vital defence industrial sectors, assesses contract-awarding procedures, examines the transparency of government's acquisition plans and intentions, discusses government's approaches to measuring success and looks at the partial-dependence on industry during operations.

A prominent theme of the past thirty years has been the increasing readiness of successive governments in the UK, the US and Germany to rely on the private sector for the supply of defence goods and services – to act as a customer, rather than a producer, for much of what is associated with defence capabilities. This trend has fundamentally affected the government–industry relationship in the defence sector. The functions fulfilled by the private sector go far beyond the mere provision of military equipment and materiel. In the equipment domain, it includes the full spectrum of through-life services, including the maintenance, repair and upgrade of kit. In addition, the private sector provides various kinds of people-support functions to the armed forces, including construction, laundry and catering services. In some countries, industry is also entrusted with protection and guarding tasks.

All of these services are provided by industry not only at home, but increasingly on deployed operations in the form of contractor support to operations (CSO). Moreover, industry plays a crucial role as a consultant to the government in its efforts to mature its capabilities as an 'intelligent customer'. Driven by financial imperatives resulting from increasing budgetary pressures, new operational requirements posed by the strategic environment and the changing nature of the private sector's involvement in defence, the British, US and German governments have substantially developed their roles as customers of the defence industry over the last decade.

Functions Passed from Government to Industry

The United Kingdom

In the UK, one manifestation of the trend towards relying on the private sector for the production of much of its defence capability was the privatisation of state-owned entities, either by the creation of joint stock companies or by inducing existing private firms to take over government assets. Under former Prime Minister Thatcher, British Aerospace and Rolls-Royce became joint stock companies, while British Shipbuilders was broken up into a number of privately owned firms. The state arsenals (Royal Ordnance and the Royal Dockyards) were treated slightly differently, with the former being auctioned to British Aerospace in 1987 and the latter initially turned into government-owned, company-operated facilities, as were the nuclear-weapons sites at Aldermaston and Burghfield. Only after 1997, under the Labour government, were the surviving dockyards at Devonport, Portsmouth and Rosyth fully privatised. Today, the only remaining state-owned defence industrial organisation is the Defence Support Group (DSG), which has the capacity to repair and even build armoured vehicles. In what might seem to some a strange arrangement, the DSG is often contracted to do work by the design authority for vehicles, which is often BAE Systems. In 2013, it was expected that the General Dynamics Scout fighting vehicle would be built in the UK by the DSG, with the coalition government simultaneously exploring possibilities for the DSG's privatisation since the 2010 Strategic Defence and Security Review (SDSR).[1]

UK governments have also been prominent in their readiness to act as a buyer of services, rather than as a purchaser of finished goods. They have regularly sought to outsource to the private sector many tasks that do not involve significant capital investment by the firms involved. These arrangements are often called public–private partnerships (PPPs), while private finance initiatives (PFIs) involve investment by firms in significant capital assets of the sort that in previous years would have been bought by the Ministry of Defence (MoD) itself.

By 2010, the MoD was buying a wide range of services from the private sector, rather than doing the work itself. On almost all of its sites, the grass was cut, the buildings maintained and the meals prepared by companies hired under facilities-management contracts. In addition, some training exercises were being designed and organised by private firms, and some training grounds were privately managed. Companies and universities were also being hired to deliver training and education, including equipment instruction.

---

[1] See Andrew Chuter, 'Britain Inks Deal with General Dynamics for Armored Scout Vehicles', *Defense News*, 1 July 2010; Andrew Chuter, 'Sale of UK Repair Company Hits Snag Over Intellectual Property Rights', *Defense News*, 29 July 2013.

The MoD had become a major user of PFIs especially, but not exclusively, for infrastructure projects. Companies were paid on a monthly basis to build and operate a large number of buildings and other infrastructure elements. More sensitively, in terms of operational needs, the MoD's satellite-communications requirements were being met (largely) by the Skynet V network of satellites and ground stations built and supported by Astrium (an element of EADS). The heavy-equipment transporter fleet, roll-on roll-off ferry fleet, and Voyager fleet of tanker and transport aircraft were also provided under PFI arrangements. PFIs could be viewed as contracting for almost all elements of capability, with the MoD putting in minimum internal elements. By 2010, the MoD was paying more than £1 billion a year in PFI service charges.[2] Of course, the move to PFIs was part of a general global trend away from contracting mainly for equipment and parts supply towards contracting for whole-life services.

For equipment, the traditional model, still in place in the late 1990s, saw a contract placed with industry for the delivery of completed systems, normally along with a set of spare parts expected to last about two years. Once a piece of equipment moved to the in-service date, either the prime contractor or a mixture of prime and sub-system contractors were placed under contract for a specific time to deliver spares according to demand.

There were at least three drawbacks to such arrangements. One was that, especially for novel systems, it was difficult to predict the requirement for spare parts, which meant that the initial package rarely matched the replacements needed. Secondly, there was no incentive for contractors to provide more reliable equipment, since they could profit by providing spare parts alone. Finally, equipment which did prove reliable could leave procurement teams with money left over towards the end of the financial year, creating the temptation to use it on (unnecessary) spares, rather than lose it back to the centre of the MoD or the Treasury.

In light of these considerations, the MoD turned to 'contracting for availability', where the contractor was required to generate a specified number of systems available for use at a given time. Such contracts were meant to incentivise the firm to invest in engineering changes that would make equipment more reliable, giving businesses a reduced risk of contract-compliance failure and less work to do. Given they could involve costly design and engineering work – that is, company investments that would take time to recoup – contracts for availability tended to be multi-year (eight or more years) in duration. The emphasis on contractor

---

[2] Defence Analytic Services Agency (DASA), *Defence Statistics 2011*, Table 1.3, 28 September 2011. See also: Ministry of Defence, *Annual Report and Accounts 2010–11* (London: The Stationery Office, 2011), pp. 150–52.

involvement in equipment support continued to grow after 2010 under the coalition government.[3]

The protracted UK military operations in the Balkans, Iraq and Afghanistan also stimulated the MoD's readiness to be a customer for, rather than an in-house generator of, services in military theatres, mainly, but not exclusively, for the support of troops. Such services included, for example, cooking, cleaning, and the provision of private communications, accommodation and maintenance. In the Balkans in the 1990s, the British armed forces became increasingly reliant on private contractors, which led to the Permanent Joint Headquarters (PJHQ) concluding a contractor logistics (CONLOG) contract with Kellogg, Brown and Root (KBR), under which planning for contractors' roles would be built into UK military preparations.

With this increasing reliance on contractors, small groups of equipment manufacturers' staff were deployed to theatre to provide advice and support. Whilst the UK military avoided using private security guards for the protection of UK military personnel, private contractors were used extensively for the often risky tasks of logistics transport and its protection.

Andrew Higginson estimates that UK CSO expenditure for 2010 was around £2.6 billion.[4] The net additional costs of UK operations in Afghanistan (£3.8 billion) and Iraq (£95 million) in the fiscal year 2010–11 – which ended on 31 March 2011 – came to around £3.9 billion, suggesting, by best estimate, that CSO expenditure accounted for at least 60 per cent of the UK's overseas operational defence sustainment effort in 2010.[5] On Operation *Herrick*, the number of companies supporting the UK in theatre rose from twenty-two (with 2,030 employees) in July 2008 to sixty-seven (with 4,867 employees) in July 2010.[6] In Afghanistan, UK contractors comprise around 40 per cent of the MoD's total workforce,

---

3 See, for instance, B Goodlad, 'UK MoD Seeks Industry Support for Vehicles', *Jane's Defence Weekly*, 27 April 2011.

4 Andrew Higginson, 'Contractor Support to Operations (CSO) – Proactive or Reactive Support?', *RUSI Defence Systems* (Vol. 13, No. 2, October 2010), p. 16.

5 See MoD, *Annual Report and Accounts 2010–11*, p. 47.

6 UK House of Commons Defence Committee, 'Written Evidence from the MoD', *Operations in Afghanistan, Fourth Report of Session 2010–12– Volume I*, HC 554 (London: The Stationery Office, 2011), pp. 90–91. It should be stressed that the authors regard these official figures on CSO spending and personnel to be conservative estimates. The authors' interviews with senior CSO experts revealed a general deficit in the government's ability to provide accurate, up-to-date figures on CSO. Therefore, although the existing official figures on CSO, as well as figures provided by analysts and commentators, present a general trend, they should be treated with caution.

according to figures from the Office of the US Deputy Assistant Secretary of Defense (Program Support).[7]

The 2012 commitment to restructure the British Army down to 82,000 personnel will likely mean that the UK military will remain a significant user of private-sector services on any extended deployed operations. However, as with the US, there is the unresolved issue of the demarcation of activities which must remain in governmental military hands. The cumulative effect of the readiness to rely on the private sector was that, by 2012, the MoD was spending around 60 per cent of its money with the private sector, which arguably should therefore be providing at least that percentage of the elements of required capability.

Clearly, this could not take place within a strictly contract-defined, transactional framework, reflecting the rather adversarial government–industry relations that appeared to inspire much of the emphasis on competitive contracting. While competitive tendering continues to be used often for the selection of a contractor, once the contract is in place the MoD is normally ready to talk in terms of 'partnering' with its suppliers, in the mutual interest of both. The precise meaning of 'partnering with suppliers' continues to be debated and to evolve, but it certainly means that MoD–industry relations were not seen as zero-sum – recognising, that is, that some events and behaviours could be in the interest of both parties.

Much of the underpinning thought justifying the MoD move towards becoming a decider and a customer rather than a provider lay in the tenets of New Public Management, which held that the private sector, when properly incentivised and controlled, would normally be more efficient and effective than the public sector. The Conservative–Liberal Democrat coalition government seemingly has held to these expectations even more firmly than its Labour predecessor. Within defence, it has set about contracting with a private company to manage the MoD's estate in the Defence Infrastructure Organisation, and to find a firm to provide human-resources and other back-office functions within a Defence Business Services organisation. Most controversially, it has also explored contracting out the contracting process itself by turning much of the Defence Equipment and Support (DE&S) organisation into a government-owned, contractor-operated (GOCO) organisation.[8]

---

[7] Office of the US Deputy Assistant Secretary of Defense (Program Support), 'US/UK Force Generation Analysis – Sustainment Strategies: Use of Contractors to Support Operations (Collaborative Element 6)', 9 October 2012, p. 6, <http://www.acq.osd.mil/log/PS/ocs/multi-national/US-UK_CE6_final_9Oct2012.docx>, accessed 8 October 2013.

[8] For a critical analysis of this possibility, see RUSI Acquisition Focus Group, 'The Defence Materiel Strategy and the GOCO Proposal for Abbey Wood', RUSI Briefing Paper, July 2012.

In the UK, there is no legal or formal policy stipulation preventing the outsourcing of 'inherently governmental' tasks, although the Ministry of Defence has recognised that such a category exists.[9] The British government has been ready to entrust almost anything except front-line operations to the private sector.

The United States

In the United States, the defence industry represents a major part of the overall US economy, but it is not a normal market. There is one single buyer, in the form of the Department of Defense (DoD) and a small group of major corporate suppliers – the prime contractors – that essentially form oligopolies in each sub-sector of defence. Moreover, it is a uniquely structured market, in that the government – as the sole buyer and regulator – plans and controls the conditions that should lead to an efficient, effective and responsive industrial structure.[10] The purpose of this is to satisfy the operational needs of defence, meet the expectations of taxpayers in terms of affordability and efficiency, and ensure successful operation within the laws of the United States.

Free-market economists in the US would prefer a set of conditions whereby there are multiple suppliers and buyers within a market; freedom of trade and movement of scarce resources; and no long-term barriers to market entry, all choreographed by the hidden hand of the pricing mechanism. The US defence and national-security market is, of course, not like this. Instead, the US government creates the conditions of economic and operational performance through the careful management of a regulated economic souk of a small number of prime contractors, sitting amongst many thousands of small, niche suppliers, with only one national buyer. An understanding of what government entrusts to industry can only really occur once this simple economic picture is recognised.

Likewise, the US need for military equipment and services is derived from specific historical conditions; so many features of today's defence industries and service businesses possess a clear historical antecedence. Elements of the research literature in this area point to a number of such considerations, which frame the US government's reliance upon its

---

[9] See Ministry of Defence, 'Better Defence Acquisition: Improving How we Procure and Support Defence Equipment', 7 June 2013, p. 17, <https://www.gov.uk/government/uploads/system/uploads/attachment_data/file/206032/20130610_WP_Better_Def_Acquisition_screen_final.pdf> accessed 15 October 2013, cited and discussed in Trevor Taylor and John Louth, 'What the Government Must Do in Defence Procurement', RUSI Briefing Paper, September 2013, <http://www.rusi.org/downloads/assets/GOCO.pdf> accessed 15 October 2013.

[10] Jacques S Gansler, *Democracy's Arsenal: Creating a Twenty-First-Century Defense Industry* (Cambridge, MA: MIT Press, 2011), p. 9.

historical base.[11] The first of these is that, historically, the US has built up its defence production in response to the requirements of a specific war, or to meet a perceived set of threats or hazards. This production has been generated through procurement programmes managed and delivered by industry, predominantly from within the private sector (defence businesses and facilities owned by the government are addressed below), but nurtured and steered by the requirements of government.[12] Politicians and officials in the US have therefore traditionally entrusted to industry the responsibility of developing and populating the national arsenal to further the national interest.

However, the volume and scale of that national armoury has fluctuated with variations in defence spending. As a general rule, the US government and population have enjoyed a peace dividend following sustained periods of conflict, either real or virtual, since at least the end of the Second World War. For example, in 1945, defence spending in the US was in excess of $720 billion.[13] This dropped in 1947 to just under $150 billion, as the US economy reverted to a peacetime footing. The Korean War prompted an increase in spending to $650 billion, before dropping to below $400 billion towards the end of that decade, as hostilities came to an end. The Vietnam War of the 1960s and early 1970s added another $100 billion per annum to the defence budget, which declined once more during the fading years of the Cold War. It rose again to levels of expenditure close to those of 1945 as the US engaged in Afghanistan and Iraq following the 2001 attacks on Washington and New York.

With each period of fluctuation in the defence budget and annual expenditure, the private sector within the US has responded to fill the national armoury on behalf of the government, and to meet the operational requirements of the day and future military ambitions. Unequivocally, the US defence and national-security industry is trusted by the government to deliver the equipment and services necessary for their national military capabilities.

The second consideration concerns a reconciliation of the tension between the unique defence market, with its continued need for, and history of, centralised industrial planning, and the US's strong belief in the power of the free market.[14] During the Second World War, centralised planning was undertaken by the Office of War Mobilization, which was

[11] John M Blair, *Economic Concentration: Structure, Behavior and Public Policy* (New York, NY: Harcourt Brace Jovanovich, 1972).

[12] Michael M Dunn, 'The US Defense Industrial Base: Past, Present and Future Challenges', paper presented at the US Industrial College of the Armed Forces, 2 June 2005.

[13] The US dollar values quoted in this section are in 2012 prices.

[14] See Murray Weidenbaum, *The Economics of Peacetime Defence* (New York, NY: Basic Books, 1974).

responsible for shifting the American peacetime, free-market economy to a wartime, part-planned economy. The Office of Defense Mobilization assumed the role in time for the Korean War, with the head of department a presidential appointment. As the budget shrank post-Korea, the status and responsibilities of the office were reduced, finally being superseded in 1991 by the Emergency Mobilization Division of the Federal Emergency Management Agency (FEMA). An established Executive Order states that FEMA holds the responsibility for industrial planning for defence, as a prime function of the Department of Homeland Security.[15] However, the health and competitiveness of the defence industry in the US remains a significant concern in the DoD and the single services.

Moreover, the US Defense Production Act, which dates from 1950, empowers the DoD to plan and manage potential production surges to meet operational demands. The president has the authority, under the act, to seize and allocate critical materials to defence, and to demand specific production-output levels from defence manufacturers registered in the US. Critically, the on-shore defence industry is entrusted by government to respond to a presidential order to, for example, surge production.

The third feature is the important recognition that there is no single, homogenised defence industrial base in the US. Rather, aircraft design and construction, land-vehicle manufacture, ship-building and munitions are all different, with divergent economic histories. Until the Second World War, much defence equipment was drawn from commercial industries, converted to war production when necessary, but government was keen to revert to civilian, commercial products when the conflict was at an end. In contrast, from 1945, a specialised defence industry grew in response to increased technological demands, including fire-control and propulsion systems. Whilst the civilian commercial sector today is equally driven by technology, with both defence equipment and civilian consumables sharing many sub-systems and parts, most defence goods still come from defence facilities. The US government has driven, and takes comfort from, this separation of defence from the wider market.

Germany

Like the UK and the US, the German government has entrusted various defence functions to the private sector, in the form of 'conventional procurement' and 'privatisation efforts'. Conventional procurement encompasses products and services procured under the Bundeswehr's Customer Product Management (CPM)[16] system, operated by the German

[15] Executive Order Number 12656.

[16] BMVg, 'Customer Product Management (amended): Procedures for Requirement Identification, Procurement, and In-Service Support in the Bundeswehr', Ref. No. 79-01-01, 12 November 2012.

Ministry of Defence (BMVg) Equipment, Information Technology and In-Service Support Directorate (AIN).[17] The CPM is the BMVg's procedure for determining and meeting Bundeswehr demands. It aims to acquire and maintain the required capabilities through a timely and economic supply of operational products and services. The CPM is the reference document for the conventional procurement of military equipment. As detailed later, less conventional privatisation arrangements are governed by a Framework Agreement on Innovation, Investment and Cost-Effectiveness in the Bundeswehr.

The Customer Product Management System: The newly revised version of the CPM, valid since January 2013, reflects the reorganisation of the German defence-acquisition system, which is part of the overall reform process of the German federal armed forces.[18] The reform of the German defence acquisition system saw a revision of procurement and in-service processes, which were separated into three phases: analysis, realisation and 'in-service' (see Figure 1). These reforms will have significant implications for the BMVg as a customer in the area of conventional procurement.

The German government's general approach to privatisation in the defence sector is described by the White Paper 2006 on German Security Policy and the Future of the Bundeswehr:[19]

> The Bundeswehr will consistently concentrate on its core tasks. Cooperation with trade and industry on service tasks, extending as far as the outsourcing of complete task packages that the private sector can provide more economically, is being pursued further. This will ease the work burden of the Armed Forces, boost cost efficiency, and reduce operating costs and tied-up capital. Private investor capital will be mobilised, new sources of revenue opened up and opportunities to strengthen investments for the Bundeswehr thereby created.

The majority of the BMVg's privatisation efforts have so far taken place in the area of the Defence Administration Organisation – a legally and organisationally independent body in the Bundeswehr. Reflecting upon those efforts, it could be argued that the explicit separation of the Bundeswehr into the armed forces and the Defence Administration

[17] Ulrich Petersohn, 'Outsourcing the Big Stick: The Consequences of Using Private Military Companies', Working Paper Series No. 08-0129, Weatherhead Center for International Affairs, Harvard University, p. 13.
[18] See BMVg, 'Die Neuausrichtung der Bundeswehr: Nationale Interessen wahren – Internationale Verantwortung übernehmen – Sicherheit gemeinsam gestalten', Broschüre zur Neuausrichtung, Zweite, vollständig aktualisierte Auflage, March 2013, pp. 92–95.
[19] BMVg, 'White Paper 2006 on German Security Policy and the Future of the Bundeswehr', October 2006, pp. 62–63.

Figure 1: Germany's Procurement and In-Service Process.

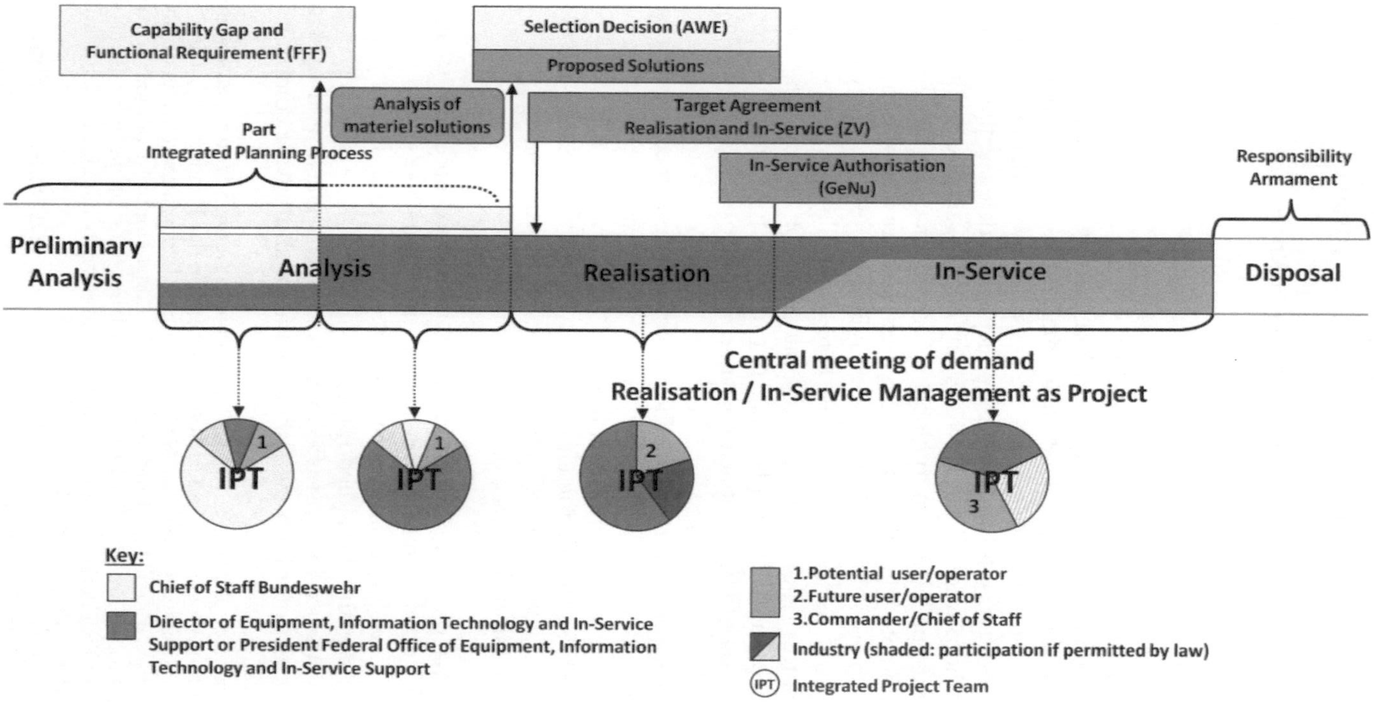

Source: BMVg 2013, translation by the authors.

Organisation has allowed the BMVg to take a proactive but sensible approach to the privatisation of services, as governed by the German constitution.

The BMVg's doctrine does not offer a single, precise definition of core and non-core military capabilities. Therefore, a rough distinction follows from the references made by multiple doctrinal documents, including the White Paper 2006 on German Security Policy and the Future of the Bundeswehr, the Conception of the Bundeswehr (KdB) and Special Conceptions (TKs) of the Bundeswehr.[20]

According to these documents, core military capabilities include five categories of functions: operational armed forces, command and control (C2), global reconnaissance, the strategic ability to deploy troops, and maintenance tasks. It is the BMVg's policy not to delegate these functional areas to the private sector. However, in the past, significant Bundeswehr capability gaps in relation to global reconnaissance and strategic deployability prompted the BMVg to obtain these services from the private sector as a temporary solution.[21] Moreover, in theatre, the BMVg may choose to outsource some elements of its core capabilities when operational circumstances allow. However, a minimum of organic capability is maintained and back-up plans are put in place to ensure the provision of service should the contractor arrangement falter.

Procurement outside the CPM: The Framework Agreement on Innovation, Investment and Cost-Effectiveness in the Bundeswehr,[22] signed by industry and BMVg representatives in 1999, serves as the basis for the Bundeswehr's privatisation efforts for services not covered by the CPM. It outlines the aims of the resulting co-operation between the two, specific to the German defence sector, as follows:

- To make use of German industry's innovation and skills to increase the German armed forces' investment possibilities
- To consolidate and perpetuate the innovative capacity of both industry and the armed forces
- To enhance the cost-effectiveness of procurement and operational processes for both the public contractor and private companies

---

[20] See, for example, BMVg, 'White Paper 2006 on German Security Policy and the Future of the Bundeswehr' and BMVg, 'Konzeption der Bundeswehr', 1 July 2013.
[21] As an illustration, consider NATO's Strategic Airlift Interim Solution (SALIS) agreement. As a member of the SALIS multinational airlift consortium, Germany has access to six Antonov An-124 transport planes for strategic airlift. See NATO, 'Strategic Airlift Interim Solution' (SALIS)', <http://www.nato.int/cps/en/natolive/topics_50106.htm>, accessed 14 October 2013.
[22] BMVg, 'Rahmenvertrag Innovation, Investition und Wirtschaftlichkeit in der Bundeswehr', 15 December 1999.

- To increase the effective use of resources in order to open up new investment possibilities.

Although more than a decade old, these targets are still valid and function as the foundation of the Bundeswehr's privatisation efforts, which can be distinguished according to three categories: formal, functional and materiel privatisation.

In the case of formal privatisation, a public body operates under private law while the state maintains full ownership. A prime example of such an 'in-house company' is the Gesellschaft für Entwicklung, Beschaffung und Betrieb (g.e.b.b. – the 'Company for Development, Procurement and Operation'). The g.e.b.b. acts as the core enabler of the BMVg's privatisation efforts. Established in 2000 by the BMVg – its single shareholder – it advises the BMVg as an in-house consultancy on the reorganisation of civilian services and all questions of efficiency. The g.e.b.b. operates independently and on its own initiative, and has the scope and freedom to develop privatisation projects under its own responsibility. Its aim is to support the Bundeswehr in fulfilling its core military tasks by relieving it of non-military service tasks.

Functional privatisation describes the foundation of PPP companies – a focus of the g.e.b.b.'s endeavours since its establishment. Both the public and private sector hold shares of the joint company. In a PPP company, the industrial partner is usually the majority shareholder, reflecting the government's intention to transfer the main commercial risk into the private sector. As an illustration, consider LH Bundeswehr Bekleidungsgesellschaft (LHBw), which was launched in August 2002. It is responsible for the purchase, storage and distribution of all Bundeswehr clothing. The BMVg holds 25.1 per cent of the company and the LHBw holds the remaining 74.9 per cent.[23]

Materiel privatisation features a full delegation of the provision of a service from the state to the private sector. This type of privatisation – essentially outsourcing – occurs only very rarely, as the level of governmental influence beyond explicit contractual arrangements is severely limited. Examples of outsourcing solutions can be found in the area of transportation and infrastructure services, such as the sea transportation of containers and military equipment, the operation of a water-bottling facility and the maintenance of an incineration site for clinical waste in theatre.

[23] LHBw is a joint company owned by Lion Apparel Deutschland and Hellmann Worldwide Logistics, each holding 50 per cent. See BMVg, 'Wirtschaftlichkeit und Projektmanagement', Bundesakademie für Wehrverwaltung und Wehrtechnik, Mannheim, Modul 3, A/GT-L:PM/AWP 3 - Kurs 02/12, 8 August 2012, p. 21.

The general decision to privatise a service formally provided by the state, as well as the selection of the specific privatisation path (formal, functional or materiel), is based on a range of factors among which cost-effectiveness is essential.

Given the above, there are a number of distinct characteristics to consider when functions are passed from the public sector to defence industry. The first is that governments now seem to favour contracting for services or capabilities, rather than just for a physical product. Even large capital items now come with upgrade and maintenance schedules, and options for integration and future disposal. Second, government seems to associate industry with efficiency and innovation, though the source of that innovation can be contested. It may be asked, for example, whether smaller businesses innovate, or whether this is the sole preserve of the large, traditional, original equipment manufacturers (OEMs). Mariana Mazzucato argues authoritatively that state funding and support for innovation has had a positive impact and her work includes a case that demonstrates how Apple has made great use of defence-funded technologies.[24] Lastly, even functions associated with the front line and operations can now be outsourced to industry. This recasting of the 'inherently governmental' is a theme across advanced industrial states, driven by both politics and notions of effective programme management, pointing towards the hegemony of the private sector.

Text Box 4: Defining 'Cost Effectiveness'.

> 'The principle of cost effectiveness stipulates that the optimum relation between the intended purpose and the funds (resources) used must be aimed at. This comprises the principle of economy and the principle of profitability. The principle of economy (minimum principle) calls for the achievement of a certain result with the least possible expenditure of resources. The principle of profitability (maximum principle) calls for the achievement of the best possible result with a certain expenditure of resources. When determining and meeting the Bundeswehr demand, the factors determining the expenditure shall be continuously assessed in an iterative process with the aim of optimizing the cost/benefit ratio.'

Source: BMVg, CPM 2007, p. 35.

Vital Industries?

Given the private sector's crucial role in sustaining national defence efforts – a core function of the state – governments tend to perceive some defence industrial sectors as vital to their defence capability. This section therefore

[24] See Mariana Mazzucato, *The Entrepreneurial State: Debunking Public vs. Private Sector Myths* (London: Anthem Press, 2013).

addresses the areas, if any, in which the British, US and German governments recognise that they have to rely on on-shore domestic defence industrial capabilities and cannot outsource to the private sector, since they regard them as inherently governmental.

The United Kingdom

In the UK, the approach to this issue has not been static. Looking back more than twenty years, to when Peter Levene (now Lord Levene) was chief of defence procurement, the government's declared policy was to run competitions which were open to international bidders from friendly countries. There was formal concern only with maintaining a UK capability in areas where external suppliers were not available for political reasons.

From the mid-1980s until 2005, UK practice in this area lacked clarity: although there was a stated readiness to allow foreign firms to bid for contracts, in practice many continued to be awarded to UK firms, not least because political considerations such as employment were often a factor in big project decisions, which were eventually taken at Cabinet level. The UK also instituted an Industrial Participation Policy, which sweetened the blow of the rejection of the airborne early warning (AEW) Nimrod and the purchase of the Sentry airborne warning and control system (AWACS). This was, in essence, an offset approach which determined that defence firms supplying goods to the UK should award defence contracts to UK businesses for sub-systems which at least equalled the value of the original contract. With regard to the AWACS contract, Boeing made specific commitments to this effect (see Chapter III).

The post-Cold War political context was of considerable importance, since the mobilisation of industry associated with the First Gulf War in 1991 and (to a lesser extent) in subsequent military operations reminded the government that the defence supply chain – preferably agile, efficient and resilient – was a significant component of defence capability in an age of surprise and sometimes protracted military operations. An excessive reliance on foreign suppliers for the adaption and sustainment of military capability would both entail operational risks and damage the UK's stance as a significant military player capable of independent military action.

The first sign that industrial considerations were achieving a significant place on the defence agenda came in 2002 with the publication of the Defence Industrial Policy, which contained few policy points beyond the formal and important recognition that foreign-owned firms that added significant value in the UK through research, development, production or support work would be treated by the government as British businesses. More significant was the Defence Industrial Strategy (DIS) document of 2005, which took a sector-by-sector approach and was driven by Lord Drayson as minister for defence procurement. This stressed that, for the UK to be able to use its equipment on unexpected operations, it needed an internal capacity to sustain and modify items in its inventory, which implied significant technology transfer from any external supplier. The degree of eventual UK capability to repair and adapt the F-35 Joint Strike Fighter (JSF) will be a key test of this ambition.

The government was never ready to fund the DIS; indeed, it remains a mystery why the Treasury signed up to the document when it was clear that funds would have to be used to preserve some industrial sectors. Lord Drayson left politics, apparently in frustration at the rate of progress. However, as a consequence of the DIS, some long-term arrangements were put in place regarding surface-shipbuilding (the Terms of Business Agreement involving BAE Systems and Babcock), helicopter and combat-aircraft support, and complex weapons. In the latter field, Team Complex Weapons was formed by bringing together the government, MBDA, Thales, Roxel and QinetiQ in an arrangement to ensure that the UK kept a capacity to design, develop, produce, test and support complex weapons. This is also true for the UK government's Porton Down site in providing a commercial and operational focus for industrial and governmental efforts to protect against biological, chemical and radiological threats.

The government's 2012 *National Security Through Technology* White Paper formally put an end to the sector strategies and, whilst maintaining a commitment to the UK's ability to use its forces as it saw fit, asserted a preference for buying 'off-the-shelf' from the world defence market and emphasised the centrality of affordability.[25] Nevertheless, the contractual arrangements with different companies remained in place and there was no suggestion that Team Complex Weapons, which served the government well during the Libya campaign, had ceased to operate. Finally, it is worth noting that a small industry team was set up in the MoD's Whitehall building to grapple with these issues and locate them within wider UK defence policy. If the 2012 White Paper sent a message that the concerns of the DIS were a low priority at best, the contracts placed soon afterwards for reactor and submarine work showed government acceptance that a UK nuclear-submarine industry is needed if the submarine-based deterrent is to be continued and replaced.

The United States

In contrast, the US keeps huge standing forces under arms which must be equipped and maintained. It requires significant support from, and

---

[25] See the International Institute of Strategic Studies (IISS), *The Military Balance 2011: The Annual Assessment of Global Military Capabilities and Defence Economics* (London: Routledge, 2011), pp. 41–51.

engagement with, the national industrial base, with certain sectors even recognised as being crucial to existential ideas of sovereign capability and autonomy.

A longstanding method for sustaining perceived key defence competencies is for that section of the defence industrial base to reside in the public sector. Historically, as the US developed following its declaration of independence, it generated a mixed private and public defence industrial base. During the nineteenth century, the government owned forty-seven military arsenals and six naval shipyards.[26] Equipment for the army traditionally came from public facilities, with the exception, perhaps, of small arms, which were drawn from the private sector, with businesses such as Colt acting as key suppliers to the government. Likewise, aircraft and other air systems, once added to state arsenals, have historically been the preserve of private corporations working to government demand and design requirements (though air maintenance depots were public facilities). Today, government maintenance, repair, overhaul and production facilities still consume in excess of $15 billion of the annual defence budget.[27] This includes small-arms ammunition factories, marine depots and naval shipyards, land maintenance facilities and air logistics hubs; all residing within a US defence industrial public sector directly employing some 70,000 civilians.[28]

At one level of analysis, therefore, consumables repair and overhaul organisations and logistics facilities are perceived to be so vital to the US defence effort that they have been kept within the public sector, despite an ongoing intellectual argument for privatisation. Other areas that are seen as essential to national capability and freedom of manoeuvre are cyber-technologies, capabilities associated with strategic deterrence, mission-system technologies and those for targeting and surveillance.[29] Whilst many of these competencies reside in the private sector, in companies such as Lockheed Martin, Boeing and L-3 Communications, they do so under the sponsorship, surveillance and patronage of government.

---

[26] See, for example, Marion E Bowman, 'Privatizing while Transforming', *Defense Horizons* (No. 57, July 2007), pp. 1–9.

[27] See Stephen Daggett and Pat Towell, 'FY2013 Defense Budget Request: Overview and Context', Congressional Research Service Report, 20 April 2012, <http://www.fas.org/sgp/crs/natsec/R42489.pdf>, accessed 8 October 2013.

[28] See Rupert Smith, *The Utility of Force: The Art of War in the Modern World* (London: Allen Lane, 2005).

[29] Ministry of Defence, *National Security Through Technology*, Cm 8278 (London: The Stationery Office, 2012), p. 8.

Germany

The German government's White Paper 2006 on German Security Policy and the Future of the Bundeswehr postulates the importance of indigenous defence-technology capabilities and requests that policy-makers and industry leaders mutually determine the strategic positioning of German defence technology in Europe:[30]

> A modern Bundeswehr requires an efficient and sustainable defence industry base ... It means having indigenous defence technology capabilities in order to co-shape the European integration process in the armaments sector. These will guarantee co-operability and assure an influence in the development, procurement and operation of critical military systems. Only nations with a strong defence industry have the appropriate clout in Alliance decisions ... The political leadership and industry must jointly define the strategic positioning of German defence technology in Europe. The Federal Government will do its utmost in this regard to preserve a balanced mix of defence technology, including its high-technology areas, in Germany.

Based on the 2006 White Paper, the BMVg and industry representatives tried to generate a common understanding of core national defence-technological capabilities through the 2007 BMVg–BDI Joint

Table 5: Germany's Vital National Defence-Technological Capabilities.

System Level	Sub-System Level
Space-based reconnaissance	Electronic reconnaissance/electronic warfare
Combat aircraft	ABC defence components
Transport helicopters	Explosive ordnance disposal components (landmines, weapons and improvised explosive devices)
Unmanned aerial vehicles	
Air-defence systems/ anti-aircraft/anti-artillery	
Armoured vehicles	
Tracked vehicles	
'Infantryman of the future' (IDZ)	
Submarines/autonomous submarine vehicles	
Surface combat ships	
Sea mine defence	
Modelling and simulation	
It-SystemBW (Operations)	

Source: BMVg–BDI Joint Declaration on Core National Defence Technological Capabilities, 2007.

[30] BMVg, 'White Paper 2006 on German Security Policy and the Future of the Bundeswehr', p. 63.

Declaration on Core National Defence Technological Capabilities, with the following objectives:[31]

- To support the performance and competitiveness of the German defence industry, including securing identified core industrial capabilities in the European context
- To provide the German defence industry with planning security regarding its current and future investment decisions
- To secure jobs in Germany through the sustainment of innovation forces for the promotion of cutting-edge technology
- To advance the development of a defence industrial base and a defence market which provides a level playing-field, also with regard to the freedom of businesses from state interference.

As such, the joint declaration aimed to identify 'capabilities, which, taking into account a sustainable European perspective on the basis of the necessary industrial competitiveness and assertiveness, cannot be abandoned for security policy, industrial policy, technological or defence industrial reasons'. Arguably this definition of core, national defence-technological capabilities requires a clear prioritisation of capabilities. However, as Table 5 shows, the joint declaration's final list of systems and sub-systems considered vital to national capability does not demonstrate this prioritisation and, instead, presents a more or less comprehensive list of Germany's defence industrial output.

Of course, such an approach predates the financial crisis that rocked the world in 2008. Its relevance today is, at best, questionable.

Summary

In summary, the UK, the US and Germany have all identified industrial sectors that they consider vital to their national defence efforts. Suppliers in these sectors provide critical products that are a *sine qua non* for military operations, such as ammunition, spare parts, or highly sensitive, cutting-edge technology, as in the case of cyber-capabilities and the nuclear deterrent. All three countries are therefore very reluctant to rely on foreign suppliers in these sectors and have taken measures to ring-fence them against foreign ownership. The US government, given its distinct notion of 'inherently governmental functions', is particularly protective, keeping some of these industrial capabilities either directly in the public domain or applying a strict control system on US-based suppliers. In the UK and Germany, the concept of 'inherently governmental functions' is not

[31] 'Gemeinsame Erklärung des Bundesministeriums der Verteidigung und des Ausschusses Verteidigungswirtschaft im Bundesverband der Deutschen Industrie e. V. zu Nationalen Wehrtechnischen Kernfähigkeiten', 20 November 2007.

explicitly taken into account as part of the defence industrial discourse. As a result, the UK and, to an even greater extent, Germany have so far been less clear with regard to the prioritisation of core defence industrial capabilities.

Awarding Contracts

The process of entrusting functions to the private sector in defence is governed by a combination of legal frameworks for public procurement – often with specific laws applying to defence procurement – and defence-procurement policy defining political preferences for sources of procurements (off-the-shelf, on-shore, multinational, bilateral, and so on) and type of contracts (cost-plus, fixed-costs, and so on).

The United States

As previously outlined, since 1945, competition between suppliers has been the principal method used by the DoD in its acquisition strategy, driven by the belief that competition over price and performance is the best way to generate innovation and battle-winning capabilities.[32] It is important to note, however, that the pricing mechanism within the defence and national-security sector is not at all characteristic of a free market. In the commercial, non-defence world, an increase in the supply of goods and services available in the marketplace correlates with a fall in the market price, and vice versa. In defence, however, the quantities of goods or services to be procured are typically fixed by both the government's budget and the notion of force structure. Under these conditions, firms have little economic incentive from the marketplace to reduce costs, other than for the initial acquisition competition.

Congress has recognised the benefits of competition and enacted the Competition in Contracting Act of 1984, which applies to all federal government procurement programmes.[33] Nonetheless, the DoD enjoys, under statute, many exemptions, so that a significant proportion of its procurement activity is awarded on a sole-source contractual basis.[34] There is a duality, therefore, in the US's defence-procurement model. On the one hand, there is a clear recognition of the championed benefits of competition and the workings of the pricing mechanism whilst, on the other, a practice of sector monopoly is tolerated and perhaps even

---

[32] See David M Walker, *DOD Transformation Challenges and Opportunities* (Washington, DC: GAO, November 2007).

[33] See Kate M Manuel, 'Competition in Federal Contracting: An Overview of the Legal Requirements', Congressional Research Service Report, 30 June 2011, <http://www.fas.org/sgp/crs/misc/R40516.pdf>, accessed 8 October 2013.

[34] United States Office of Management and Budget (OMB), *Report on Competitive Sourcing Results, FY 2007* (Washington, DC: OMB, 2008); Gansler, *Democracy's Arsenal*, p. 282.

promoted within defence (see Chapter III). Put simply, this is because there remain dramatic differences between the domestic commercial market and a highly regulated defence marketplace dealing in the projection of, and support to, lethal force.

In domestic markets, products are typically based on proven technology and identifiable brands that both appeal to the consumer – rather than, necessarily, the purchaser – and can be rapidly applied across the market. Within the defence and national-security sector, the technology of capital assets is usually cutting-edge, born of long-lead research-and-development (R&D) activities, but slowly applied. Likewise, within most civilian industries the barriers to entry and exit, in the mid-to-long term, rarely exist, with significant migration into and out of a particular market. With the business of defence, there are extensive barriers to entry and exit, including regulation, past contractual reputation and intellectual property. Moreover, risks and profits for commercial-market companies are invariably carried by the firm and constrained by market competition. Within defence, by contrast, risks in particular are often shared between the company and the government, though regulated by the latter.[35]

In terms of the US preference for some form of competition across the defence and national-security space, highly technologically advanced or research-led requirements tend to be handled in a specific way. The Defense Advanced Research Projects Agency (DARPA) tends to invite known businesses with a history of successful work in a particular sector or, failing that, a niche supplier possessing a specific, advanced technology to discuss informally the defence requirement with officials at a pre-tender stage. Thereafter, on the basis of these discussions, DARPA limits the competitive phase to two or three firms which have impressed during the informal discussions. The down-selected successful business competes on price but also on know-how and technical ability, which has to be proven within articulated performance parameters during a prototype phase of the programme before production funds are released by government. In this manner, some form of competition does feature across highly technical programmes in the US, but the procurement quickly moves to value-adding features of project management, technical design and integration, rather than energy and cash-sapping commercial business development and organisational rivalry.[36]

There is a difference, of course, between full and open competition and limited competition. Competitors will attempt to be innovative and

[35] See Frederic M Scherer, *The Weapons Acquisition Process: Economic Incentives* (Cambridge, MA: Harvard University Press, 1964).

[36] James Richardson and James Roumasset, 'Sole Sourcing, Competitive Sourcing, Parallel Sourcing: Mechanisms for Supplier Performance', *Managerial and Decision Economics* (Vol. 16, No. 1, January/February 1995).

cost-aware in the latter case because there is a greater possibility of winning the tender. Research in the US suggests that open competition tends to disincentivise firms, as the high costs of generating proposals for competition mean that many businesses simply reach commercial 'no-bid' decisions. This is one reason why the DoD seems to prefer single-source tenders or limited competitions.[37]

The United Kingdom

Within the UK, the MoD's preferred procurement approach has long been to use competition, either in the form of sealed bids or a competitive dialogue process. The UK government has a broad and longstanding commitment to the claimed beneficial effects of competition. These were underlined under former Prime Minister Thatcher and then her successors. As a customer, the MoD generally has liked to be in touch with a range of suppliers and to secure their best offers on a competitive basis. Despite the euroscepticism in many parts of the Conservative Party, there has been no effort by the coalition government to damage the EU's Defence and Security Procurement Directive,[38] with its emphasis on the centrality of pan-European competition. The headline statement of the 2012 *National Security Through Technology* White Paper was that 'wherever possible, we will seek to fulfil the UK's defence and security requirements through open competition in the domestic and global market, buying off-the-shelf where appropriate, in accordance with the policies set out in this paper'.[39]

However, the UK has experience of the fact that, in fields where orders are few and far between and the entry costs for would-be suppliers are very high, the effect of competition has been to destroy the supply base. The firms that lost a competition could not afford to maintain the relevant capability until the next opportunity and so left the defence sector, normally on a voluntary basis as a strategic choice. Partly as a

---

[37] Defence Materiel Organisation, *Defence Capability Plan 2011*, 15 May 2012.

[38] 'Directive 2009/81/EC of the European Parliament and of the Council of 13 July 2009 on the coordination of procedures for the award of certain works contracts, supply contracts and service contracts by contracting authorities or entities in the fields of defence and security, and amending Directives 2004/17/EC and 2004/18/EC with EEA relevance', *Official Journal* (L 216, 20 August 2009), pp. 76–136, <http://eur-lex.europa.eu/LexUriServ/LexUriServ.do?uri=OJ:L:2009:216:0076:01:EN:HTML>, accessed 8 October 2013.

[39] See DASA, *UK Defence Statistics Compendium 2011*, Table 1.15, <http://www.dasa.mod.uk/index.php/publications/UK-defence-statistics-compendium/2011>, accessed 8 October 2013; DASA, *UK Defence Statistics Compendium 2010*, Table 1.15, <http://www.dasa.mod.uk/index.php/publications/UK-defence-statistics-compendium/2010>, accessed 8 October 2013; DASA, *UK Defence Statistics Compendium 2009*, Table 1.5, <http://www.dasa.mod.uk/index.php/publications/UK-defence-statistics-compendium/2009>, accessed 8 October 2013.

consequence, today the UK has just one major domestic firm for complex weapons, nuclear submarines, nuclear-propulsion systems, gas turbines, surface ships, manned and unmanned combat aircraft, and airborne radars. The basic government response to this phenomenon has often been to open up the UK market to firms from around the world, but this is not a unique situation. Even in the US, defence industrial consolidation has proceeded to the point where the US government has virtually no choice of national supplier.

Furthermore, defence acquisition provides different sorts of value to different stakeholders, and the employment implications of defence choices can be significant. These were particularly apparent in the UK decisions to sustain a capability to build surface warships. The Terms of Business Agreement (TOBA) linking the MoD, BAE Systems and Babcock at the strategic level was about maintaining the capacity to design, build and support surface warships in the UK, but it surely would have had less appeal to government had not so many jobs in Scotland been involved.

As previously noted, successive governments have been keen to transfer public assets to the private sector and, in some cases, this has required the provision of contracts without competition as a means of inducing companies to pay to take on the new body.[40]

A lack of competition may also occur due to other defence requirements. First, the UK operates units of special forces whose procurement practices are understandably opaque, although they are understood to have close relations with a number of suppliers. Perhaps it would be reasonable to conclude that their practices are essentially pragmatic and little influenced by any doctrinaire commitment to competition as such.

Second, having been engaged in difficult military operations since 2003, the UK has made extensive use of Urgent Operational Requirements (UORs) in which speed of delivery has normally been the key element, while competitions, of course, take time.

The issue of time has also highlighted the challenge of empirically measuring the benefits of competition in major defence development projects. It is not normally possible to experiment by running two procurement strategies at the same time for the same project. Also, for reasons noted in the next section, the MoD may end up paying a very different price for a problematic project than that spelled out in the original competitive bid offer.

---

[40] Two examples demonstrating this are the ammunition contracts signed with BAE Systems in connection with the privatisation of Royal Ordnance and the long-term arrangements for test sites agreed with QinetiQ as part of the break-up of the former Defence Evaluation and Research Agency.

Finally, in just a few areas, the UK has recognised that it should have (and can, for the moment, afford) a national industrial capability, which must be kept alive by orders. Nuclear warheads and nuclear submarines, including their propulsion systems and complex weapons, enjoy this status, with the MoD needing to place work so that the firms involved can maintain the ability to design, develop, build, test and support the equipment involved.

Reflecting the emphasis placed on competitive tendering from the mid-1980s, the proportion – by value and number – of contracts awarded by competition tended to rise until the mid-1990s, although there was considerable variation from year to year, as data from the Defence Analytical Services Agency (DASA) show. Since then, the share comprising what are essentially cost-plus contracts has tended to increase to around 40 per cent of the total by value, and contracts awarded by competition had fallen to 48 per cent by value and 26 per cent by number in 2010–11.[41]

Germany

The German legal framework for the awarding of Bundeswehr contracts (see Figure 2) encompasses federal law, ordinances and administrative regulations, and complies with European legislation as set out in the EU's Defence and Security Procurement Directive.[42] National or European procedures are applied depending on the extent and type of the services required and whether the government considers vital national-security interests to be affected. The process of tendering and awarding Bundeswehr contracts is highly formalised and leaves little room for political preference.

Defence contracts are tendered at the European level for supplies and services contracts if they have an estimated value (excluding VAT) of €400,000 or above, while for construction-services contracts the estimated

[41] Gansler, *Democracy's Arsenal*, pp. 282–84.

[42] See Bundesministerium der Justiz, 'Vergabeverordnung für die Bereiche Verteidigung und Sicherheit (VsVgV)', 12 July 2012, <http://www.gesetze-im-internet.de/vsvgv/>, accessed 8 October 2013; Bundesministerium der Justiz, 'Verordnung über die Vergabe öffentlicher Aufträge (VgV)', 9 January 2001, <http://www.gesetze-im-internet.de/vgv_2001/>, accessed 8 October 2013; 'Bekanntmachung der Vergabe- und Vertragsordnung für Leistungen – Teil A (VOL/A)', 20 November 2009, <http://www.bmwi.de/BMWi/Redaktion/PDF/Gesetz/verdingungsordnung-fuer-leistungen-vol-a-2009>, accessed 8 October 2013; 'Vergabe- und Vertragsordnung für Bauleistungen 2012 Teil A (VOB/A)', 23 August 2012, <http://www.bmvbs.de/SharedDocs/DE/Anlage/BauenUndWohnen/vob_2012_a.pdf >, accessed 8 October 2013; 'Bekanntmachung der Vergabeordnung für freiberufliche Leistungen (VOF)', 18 November 2009, <http://www.bmwi.de/BMWi/Redaktion/PDF/Gesetz/vergabeordnung-fuer-freiberufliche-leistungen-vof.pdf>, accessed 8 October 2013.

Figure 2: Application of Public Procurement Law for Bundeswehr Contracts.

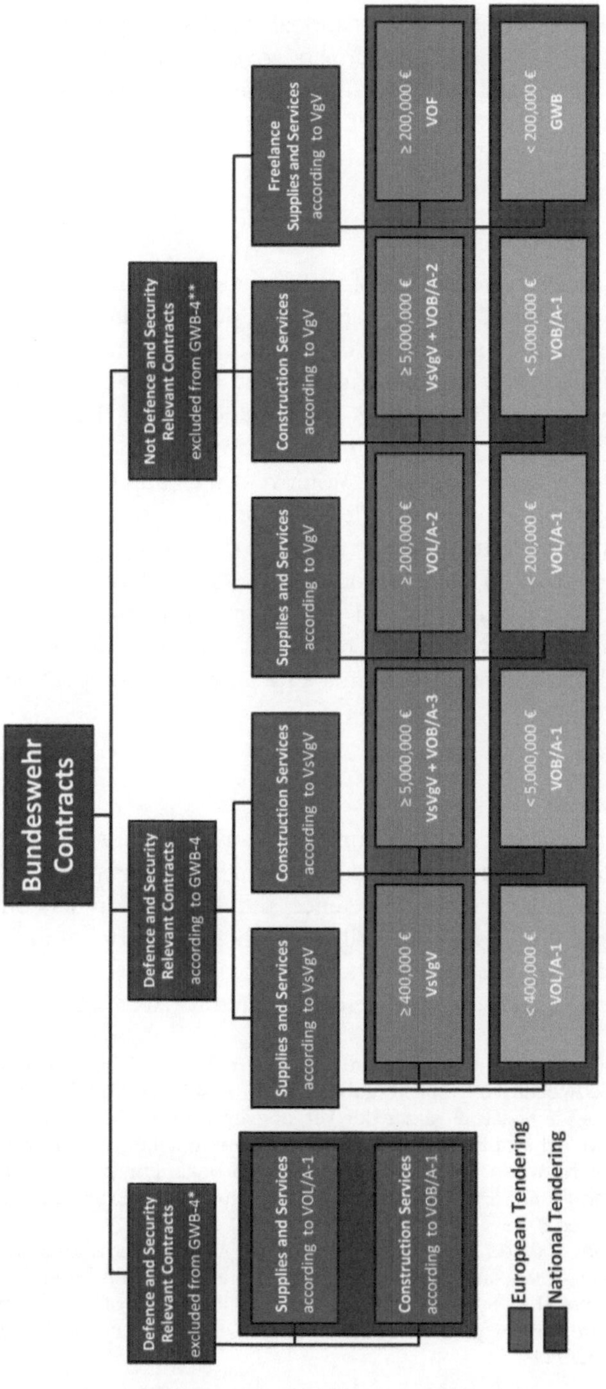

*because they do not conform with § 100 (6) GWB

**either because they do not conform with § 99 (7) GWB or are subject to §§ 100 (excl. (6)), 100a, 100b, 100c GWB

Source: The authors, 2013.

value (excluding VAT) must be €5 million or above. If the estimated values are below these thresholds, a national tendering process may be undertaken.[43] However, if the German government believes that vital national-security interests will be affected by a defence- and security-relevant contract, it may exclude it from European tendering processes regardless of these thresholds.[44]

Summary

Whereas the general preference of policy-makers in all three countries favours competitive tendering of contracts in defence, the above analysis clearly highlights that the awarding of contracts with limited or no competition at all is a consistent feature in the UK, the US and Germany. This common characteristic reflects the specifics of the defence market, which often has only a limited number of suppliers or, indeed, just one supplier for a specific product. Moreover, the ring-fencing of core capabilities within the domestic defence industry for the purposes of national security is another reason for limited or non-competitive contract awarding. Therefore, all three countries face the challenge of advancing their single-source contracting and long-term partnering skills. Furthermore, they have to balance the often legitimate need for non-competitive tendering with their political preference for increased competition in the defence market.

Transparency

The level of transparency of government acquisition plans and intentions is of particular importance to industry, which depends on a reliable understanding of government's defence requirements in its long-term strategic corporate planning, as well as to allies and partners in terms of their ability to co-ordinate and, wherever possible, streamline capability planning processes and consolidate defence industrial capabilities.

---

[43] See Bundesministerium für Wirtschaft und Technologie, 'Bekanntmachung der geltenden EU-Schwellenwerte für Vergaben in den Bereichen Verteidigung und Sicherheit gemäß § 1 Absatz 2 Satz 2 der Vergabeverordnung für die Bereiche Verteidigung und Sicherheit', Bundesanzeiger, Bekanntmachung, 25 July 2012, p. B2, <http://www.bmwi.de/BMWi/Redaktion/PDF/B/bekanntmachung-geltende-eu-schwellenwerte-fuer-vergaben-bereiche-verteidigung-und-sicherheit.pdf>, accessed 8 October 2013.

[44] See paragraphs 100 and 100c of the 'Gesetz gegen Wettbewerbsbeschränkungen (GWB)', 26 August 1998, <http://www.gesetze-im-internet.de/bundesrecht/gwb/gesamt.pdf>, accessed 28 September 2013, and art. 346, para. 1 of the 'Treaty of the Functioning of the European Union (TFEU)', *Official Journal of the European Union*, 2012/C 326/01.

The United Kingdom

For the UK, the transparency of the government's acquisition plans and intentions was an issue with which the Labour government struggled throughout its period in office (1997–2010). On the one hand, there was an inclination to be as open as possible about future plans, as Australia had been for some years.[45] The logic was that companies would feel more confident about investing their own money in a capability area if they were fairly sure that a relevant MoD contract would be forthcoming in the foreseeable future. Such investment would perhaps reduce the costs to the MoD, but should certainly speed up delivery after a contract has been awarded. On the other hand, there was an established concern that, if the MoD revealed how much money it had set aside for any particular project, it would result in companies bidding just under that ceiling. Also, especially in the latter years of the Labour administrations, it could be surmised that there was a reluctance to draw attention to over-commitment in the plan. Whatever the truth, the Labour government never managed to publish a version of the Equipment Plan, confining itself to what was felt to be releasable during 'industry days', arranged by individual heads of capability areas. The Conservative–Liberal Democrat government that followed stressed that it would reveal planning details, but had not done so by the summer of 2012. Its explanation was that the draft plan, with all the cuts made, had been sent to the National Audit Office (NAO) for scrutiny. It was not clear whether the NAO would state clearly that the plan was affordable, whether it would only pronounce on the MoD's processes to assess affordability, or whether it would follow some other course of action.

The UK does, however, have a track record of being transparent about many of its immediate needs to the global defence industry, and not just to British firms. Its monthly *Defence Contracts Bulletin*, which dates back to the 1990s, contains much information about likely contracts, invitations to contract and contracts awarded, although it does not include classified information. It is available to all for a modest subscription fee.[46]

From a governance perspective, there is also the matter of transparency in relation to the public. Here, a major role is played by the information provided by the MoD to the NAO, whose reports often serve as the basis for specific investigations by the House of Commons Defence and Public Accounts Committees. In short, the MoD wants to be more open about its future plans, but not quite yet.

---

[45] Currently published by BIP Solutions. See <http://www.bipsolutions.com/BusinessIntelligence/ContractInformation/MOD_DCB.html>, accessed 10 October 2013.

[46] Gansler, *Democracy's Arsenal*, p. 91.

The United States

In the US, the DoD publishes its Quadrennial Defence Review (QDR), which is an overall strategic statement of trends, intent and future requirements. The QDR articulates the direction of defence and national-security policy promoted by the president and, as such, is both a component of organisational strategy for the DoD and an expression of political will from the White House.

For example, QDR 2006, undertaken at the beginning of President George W Bush's second term in office, focused on the notion of the 'long war'. This iterated the logic of and need for lengthy military action in Iraq and Afghanistan, and large standing forces in the Middle East and elsewhere. It emphasised that American military effort would be undertaken in conflicts of occupation where the adversaries were not necessarily regular forces of rival nation states, but irregular warriors using asymmetrical and problem-solving techniques of war. Consequently, QDR 2006 detailed the sort of support and innovation the DoD required from industry to prosecute these sorts of operations successfully.

Due to the fact that such irregular warfare is characterised by relatively large, multipurpose ground forces of occupation operating from within a mixed population of civilians and insurgents, many US battle systems, such as tanks and advanced stand-off weapons, have been rendered impotent.[47] America's industrial-age weapons systems need to be supplemented by a new 'information-age' set of capabilities, with an operational emphasis on flexibility, timeliness and the immediacy of the decision to attack. QDR 2006 signalled to industry a requirement for ground robots and advanced unmanned aerial vehicles (UAVs), which were to prove highly effective in the battlespace over Afghanistan later in the decade.

Then-Defense Secretary Robert Gates picked up this theme in 2008, when he stated in a speech at West Point that the military had to shift from large, costly weapons systems designed to fight the conventional state-on-state wars of the past, to a focus on new ways of training and operating to interdict today's amorphous networks of combatants. For Gates, brutal and adaptive insurgencies and the proliferation of terrorism would lead to protracted and messy unconventional campaigns where information and flexibility drove military success rather than weapons systems.[48]

Such a public iteration of government policy and forward thinking has shaped industry's responses to the needs of its principal client. Corporations were comfortable about making R&D investment decisions based on this clear articulation of government need.[49] Budget shifts within

[47] Greg Grant, 'Gates Tells Military Services to Prepare for Unconventional Wars', *Government Executive*, 22 April 2008.
[48] Author interview with a US defence official, 30 March 2012.
[49] See IISS, *The Military Balance 2011*, p. 43.

the DoD towards the end of the decade further indicated that the future for the military was predicated on the development and deployment of UAVs and unmanned combat air vehicles (UCAVs), with Gates proposing, in the 2009 defence budget, to halt the production of F-22 aircraft to fund further UAV and UCAV development programmes.

In 2010, the Obama administration published a number of significant documents which have been used by the US military and defence industrialists to identify and plan for both the near and mid-to-long terms. This year saw the publication of QDR 2010, the 2010 National Security Strategy, the Nuclear Posture Review, the Ballistic Missile Defense Review and the Space Posture Review. These documents and activities reflect a coherent decision by the Obama administration to continue to focus on flexible problem-solving capabilities and information superiority. They prioritise a new Air-Sea Battle concept, long-range strike, space and cyberspace capabilities, and innovative, multi-skilled Special Forces. For the latter, QDR 2010 and the National Security Strategy detailed 660 special-operations teams, fully supported by three ranger battalions, as well as 165 tilt-rotor and fixed-wing mobility and fire-support primary mission aircraft. Even in the usually sensitive area of special operations, the US government seems hard-wired to be transparent and open in its intentions.

This, in part, is due to the nature of public budgeting in the US. As noted previously, the US polity is a highly balanced organism of checks and balances between an executive, Congress and the judiciary. Feeding the three constituent elements of government is a highly advanced lobbying and intelligentsia strata, and thus programmes of expenditure, although proposed by the president and his Cabinet (often derived from manifesto commitments), have to be agreed by Congress, which oversees public expenditure. Moreover, such programmes have to be viewed as lawful and 'American' in terms of the constitution. This drives transparency as officials are loath to risk programmes simply because an element of the complex governmental grid has been taken by surprise.

Germany

The general degree of transparency in the German government's defence industrial policy has improved over the last decade. As part of a more proactive information policy designed to communicate the overall reform of German security and defence policy, the BMVg – under the political leadership of the minister of defence – has outlined the basic foundations and general direction of the government's defence industrial policy.

Through a series of official, unclassified publications by the BMVg, public speeches and media appearances by the minister of defence, and the publication of external audit reports, the government has presented the structure and processes of the new German acquisition system. In addition, the BMVg plays an active part in the organisation and presentation of conferences and seminars examining the procedures of the new Equipment, Information Technology and In-Service Support (AIN) and Infrastructure, Environmental Protection and Services (IUB) Directorates in the BMVg. However, public information is usually limited to abstract statements regarding the organisational set-up of the German acquisition system, focusing on technical aspects rather than providing detailed information on the reasoning behind the actual policy-making process.

There are no unclassified reports that allow external analysts to comprehensively assess the track record of the government's defence industrial policy. In terms of privatisation, the economic feasibility studies (WUs) and performance documentation produced by the BMVg's Budget and Management Accounting Directorate are classified, prohibiting public scrutiny of privatisation projects. This lack of transparency echoes the traditionally low interest of the German public in security- and defence-policy matters.

Summary

The transparency of governmental defence acquisition plans and intentions differs quite substantially between the three countries, although the UK, the US and Germany all consider this to be an important factor in their defence industrial policies. The US government, due to its more mature ability to think strategically, has developed a comprehensive set of conceptual policy documents that sensibly informs the public and the US defence industrial base about its acquisition plans and intentions. Both the UK and, particularly, Germany are still restricted in their efforts to follow the US example in this area by the specifics of their national organisational and policy-making cultures. However, they seemed to have acknowledged this strategic gap and started to address the issue.

Success: A Governmental Narrative

The measurement of success in defence acquisition is problematic. The crucial considerations from the established perspective of the project manager, usually acting for an external client, is whether a project is delivered on time, within budget and to the required specification. The very idea of 'success' is highly subjective, as there is a clear difference between the success of military adventures and operations, when politicians and service personnel often think in terms of 'outcomes', and success in procurement, when the focus tends to shift to programme- and project-management skills and adherence to plans and budgets. Success, consequently, is a difficult idea to conceptualise let alone measure, and this challenge is present across the British, US and German political systems and defence industries, where narratives of success and failure become highly politicised.

In political terms, this has been the main concern within defence ministries, governments as a whole, parliaments – including their agents, such as the British NAO, the US GAO and the German Bundesrechnungshof – as well as national political discourses in the media. When a project runs over budget, it is then easy to reach a conclusion that money has been 'wasted'. However, time, cost and performance on initial delivery are not the whole story: if they were, St Paul's Cathedral and the Houses of Parliament in London would still be regarded as project failures. Therefore, a more comprehensive assessment of success would ask whether industry delivers defence programmes and projects on time and to cost, whilst concurrently meeting the demand for technological advantage, as well as considering whether it produces reliable solutions that are readily available domestically and on operations when and where required.

The United Kingdom

For at least the last fifteen years, British governments have sought to increase the weight of through-life considerations on defence projects, initially with regard to the total cost of operating a system in terms of fuel, parts and support. Over the past eight years, the government has sought to check whether all of the elements needed for capability, in addition to the equipment itself, have been provided in a timely and affordable way. It has developed the TEPIDOIL (Training, Equipment, Personnel, Infrastructure, Doctrine and Concepts, Organisation, Information and Logistics) framework as a capability management tool, and the NAO now delivers a brief report on these elements when it scrutinises the MoD's major projects. However, the MoD has struggled to adapt its management accounts so that the through-life costs of a system can be reliably measured, and the NAO stops monitoring the top projects as a group once they come into service. Thus the chief output against which DE&S project teams are assessed still tends to be time, cost and performance on initial delivery.

As a customer, the MoD also does not systematically seek to assess the utility of its purchases once in service. This, of course, is the real test of projects: a road built on time, to budget and to specification but which no one ever wants to use cannot be considered a success. World politics can complicate matters for force planners in the UK and elsewhere: the end of the Cold War greatly reduced the value of the Typhoon aircraft's agility. It also diminished markedly the relevance of the UK nuclear deterrent. The Warrior armoured fighting vehicle, designed to survive in a high-intensity battle against Warsaw Pact forces, happened to prove useful in the Balkans in the 1990s as a means of protecting convoys carrying supplies.

The fleet of 430-series vehicles was successfully modernised and upgraded in the time and for the money expected, but they were not relevant to Afghanistan because of their vulnerability to improvised explosive devices (IEDs).

Thus, the MoD's success as a customer tends to be judged, both internally and externally, on the basis of whether projects arrive on time, within budget and with the specified performance. On these grounds, with the current cohort of major projects well advanced in their development, the prospect of further problems appears modest.[50]

The United States

In the US, a GAO review is undertaken annually on selected defence and national-security activities. For example, the GAO discovered that the programmes encompassed within the 2007/08 portfolio of defence acquisitions had grown beyond their initial budgets by 26 per cent, and were therefore assessed as providing an unaffordable portfolio.[51] However, with development-cost growth – estimated by the GAO to be 40 per cent above initial project estimates for these same programmes – removed from the equation, the programmes were effectively performing on, or near to, plan. Such is the complexity of making judgements on multi-dimensional defence activities. Nonetheless, the issue of 'cost creep' plagues defence planners and industrialists alike. The GAO discovered that 63 per cent of programmes whose schedules were adjusted or performance criteria amended through the life of the programme generated cost increases of 72 per cent.[52] So the prevailing sense is that industry is not delivering to cost.

As for delivery on time, the average cycle time of a military product in the US – the time taken from development of goods or services to production or in-service realisation – has grown from just under seven years to nine over the last twenty years, demonstrating, for some commentators, a latent defence-market inefficiency.[53] In the case of the US Army and its Future Combat Systems programme, for example, the acquisition cycle time was planned at ninety-one months, but this grew during implementation to 139 months – a 52 per cent delay to the delivery

---

[50] GAO, *Defense Acquisitions: Assessment of Selected Weapons Programs*, GAO-08-467SP (Washington, DC: GAO, March 2008).

[51] *Ibid.*

[52] Over the same time period, the car industry in the US has reduced its product-realisation cycle time from about eight years to two. See Daniel Czelusniak, 'Defence Science Board Briefing', 12 June 1998.

[53] GAO, *Defense Acquisitions: Assessments of Selected Weapons Programs*, GAO-07-406SP (Washington, DC: GAO, March 2007).

of the programme when compared to the planned schedule.[54] The enduring perception in the US, consequently, is that industry can fail to deliver on time, as well as to budget.

Given that the US military is constantly in use on deployed operations ranging from aid to civil communities to war-fighting, the reliability and availability of its kit and systems are significant organisational and political considerations. Yet the US has not always been able to boast about mission-system reliability or availability. For example, the F-22 fighter aircraft has suffered from a critical failure every two hours on average, meaning that its availability to commanders has been only about 56 per cent.[55] With troop casualties on operations a regular occurrence, press and official reports of system shortcomings cause significant reputational damage to both defence businesses and those in government responsible for acquisition programmes.[56] This impacts upon how the defence base is judged by society.

In the logistics domain,[57] which is by far the largest sector of military contracting in the US, there are over 2,000 separate military and government logistics information systems, many of them standalone.[58] This works against active coherence and integration so that, despite the size of the annual budget, the US has significant issues with its defence logistics effort. According to Gansler, in 2008 alone there were more than half a million back orders, in the myriad logistics systems, still to find their way to the front line. The DoD, moreover, was unable to account for 50,000 shipping containers sent to operational theatres loaded with inventory, whilst more than 37 per cent of all munitions held in stock

---

54 R Jeffrey Smith, 'Obama Vows a Veto in Dispute over F-22s', *Washington Post*, 14 July 2009.

55 See, for example, Renae Merle, 'Marines Seek Fuse on Vehicle: General Dynamics Design has Problems', *Washington Post*, 17 February 2007.

56 By logistics, the authors refer to the ability of a nation to project, sustain and recover combat-ready joint forces through effective functions of transport, supply, maintenance, replenishment and support. See John Louth, 'Leadership, Industrial Licence and Logistics: The Search for Some New Thinking', *RUSI Defence Systems* (Vol. 14, No. 2, Autumn/Winter 2011), pp. 39–42.

57 US expenditure on logistics in a given year can be in excess of $172 billion, with an actively managed inventory worth $94 billion being carried on the DoD balance sheet. Over 18 million parts are ordered and delivered each year across all operating theatres, with upwards of 5 million equipment reference-code numbers held within the US inventory. Moreover, the military's logistics system comprises over 1 million military and civilian government personnel, matched by as many contractors from industry. It is an enormous and complicated undertaking. See Lou Kratz, 'Defining the Future of DoD Logistics', Lockheed Martin, Washington, DC, 2008.

58 Gansler, *Democracy's Arsenal*, pp. 217–19.

were classified as obsolete, beyond repair or unusable in some other manner.[59]

The US defence inventory comprises assets valued at $94 billion. However, below this number lurk some unpleasant realities. Of the US Air Force inventory of $18.7 billion, more than half is not required for operations and offers no utility at all to the military, representing an unacceptable charge on capital. Of this value, some $300 million of new equipment and parts delivered via the logistics system is immediately marked for disposal upon reaching the end user. By both reputation and performance, the defence logistics function in the US is profoundly suboptimal.[60]

Germany

In Germany, the Budget and Management Accounting Directorate has developed the comprehensive target and key performance indicator *Kennzahlen* (KnZ) system for all companies under BMVg ownership, which assesses companies' effectiveness and efficiency goals. Using this system, the directorate biennially generates an integrated report on all companies under BMVg ownership for the ministry's political leadership. Combining the user, contractor, budgetary and management perspectives, the classified report enables the political leadership of the BMVg to conceive the current and future level of the companies' operative effectiveness and efficiency. The key performance indicators and assessment of the projects provided by project heads are also the basis for periodic reports by the BMVg to the Budget Committee of the German federal parliament.

For conventional procurement enacted under the CPM procedure, 'success' at the end of its realisation phase is measured through 'compliance demonstration' and 'operational testing'. As part of the acceptance procedure, the contractor must prove compliance with the contract. Acceptance is based on the criteria for contractually agreed performance. If possible, an 'integrated compliance demonstration' must be sought, while clearly distinguishing between customer and contractor responsibilities. The customer's technical and user tests are to be harmonised to avoid duplication wherever possible, and to ensure the greatest possible coherence of the tests (in terms of place and time). Of course, this means that, much like in the British case, the German system

59 Author interview with a US defence official, 30 March 2012.

60 MoD Defence Equipment and Support, 'Contractor Support to Operations (CSO): Policy Overview, Joint Service Publication 567, 5th edition', Defence Council, 2009; MoD, 'Contractors on Deployed Operations (CONDO) – Processes and Requirements, Defence Standard 05-129, Issue 4', 12 March 2010; MoD, 'Contractors on Deployed Operations, DEFCON 697, Edition 12/10', 2010.

does not adequately account for the operational performance of a system once it has entered into service. A systematic, through-life measurement of success is, therefore, still missing in the conventional German procurement approach.

As an independent supreme federal authority subject only to the law, the Bundesrechnungshof, among other government departments, regularly examines the BMVg's financial management decisions, including the procurement of both products and services. Through 'management letters', it reports its findings to the BMVg, which is required to submit its comments on the audit findings and conclusions within a timeframe set by the Bundesrechnungshof. Moreover, the Bundesrechnungshof submits an annual report to both houses of the German parliament, the Bundestag and the Bundesrat, as well as to the federal government. Further, the Bundesrechnungshof may at any time submit special reports on matters of major significance to both houses of parliament and to the federal government. Despite these powers, the Bundesrechnungshof is not yet perceived amongst the general public as a high-level scrutinising body for government defence expenditure in the same way as the NAO is in the UK, for example. In the past, most of its often quite technical reports on the subject have not been picked up by the media and have been used only randomly by the parliamentarian opposition to hold the government publicly accountable.

Summary

The UK, the US and Germany therefore face the common challenge of measuring the success of products provided by industry over the whole lifecycle. Moreover, the concept of value for money in defence is not limited to the delivery of products against budgets and timelines, but also encompasses the operational usability, reliability and availability of equipment and material. As such, governments must comprehensively understand, monitor and evaluate the private sector's performance in the production, maintenance and in-service support of their products. This requires an effective departmental through-life project controlling system. Neither the UK, the US nor Germany has so far been able to establish such a system on a full scale, although, as shown above, some improvements have been made.

Operations: The Role of Industry

The United Kingdom

The UK has undertaken a significant number of military operations since the end of the Cold War. With the exception of the 2003 campaign in Iraq, all were rather difficult to predict and therefore to plan for. As a consequence, the existing forces at the onset of an operation were often not properly equipped or otherwise prepared. Reflecting this, the MoD developed special mechanisms for the procurement of UORs – goods and services needed comparatively quickly for a military operation. Under an established arrangement with the Treasury, the agreed marginal additional costs of a campaign do not need to be funded from the main defence budget, and extra money is granted first by the Treasury and then by Parliament. For equipment bought on this basis, the assumption is that it will only be needed for the campaign in question and so only its immediate support needs must be considered. Should the particular service branch decide that it wishes to keep a piece of UOR equipment after finishing a campaign, it must meet the support costs from the base defence budget.

Following the protracted campaigns in Iraq and Afghanistan, UORs can be placed into one of two broad categories. In the first are items that are recognised before the onset of the campaign as required to increase the chances of success on acceptable terms. Delivery of these items is required within a small number of months at most and the time for contracting is very limited. The need for these items occurs because they do not form part of the force elements kept at readiness by a peacetime budget for generic types of campaign: specific campaigns almost always bring their own needs. Also, government can be reluctant to fund the provision of certain war stocks, especially if this would involve significant expense or if there is confidence that the items could be procured easily from the civil economy: AA batteries being one such example.

In the second category are demands that arise because a campaign has not progressed as originally hoped, with adversaries using technologies and tactics for which UK forces are not well prepared. In Afghanistan and Iraq, UK forces developed specific needs as the campaign unfolded, not least for the identification of, and protection against, IEDs. For such requirements, the time pressures for delivery have been slightly more relaxed, perhaps extending to a year or more.

As a broad generalisation, the MoD has looked mainly to UK-based plants to fulfil its needs in the first category, sometimes expecting them to start work before a contract has been signed or modified. Industry has generally been co-operative, and for the Libya campaign of 2011 it even earned the express gratitude of government. For military campaigns, it has been in the MoD's interest that industry and government should be seen to be 'in it together', working co-operatively to protect the lives of UK forces and to enhance the chances of British military and political success. For UORs in the second category, the UK has looked much more to the global marketplace, having to take only limited account of any equipment's long-term support needs. Generally, the MoD as a customer appears to have been satisfied by the efforts of its own procurement staff and the responses of industry at home and abroad in relation to UORs. Delivery has been prompt and there have been no accusations of company exploitation of any vulnerability in the government position, at least in the public domain.

In addition to the government's experience with industry's ability to deliver quality UORs in time and on budget, as mentioned above, the private sector's role on operations also encompasses the provision of CSO services. Here, too, the overall experience can be considered positive.

The MoD has developed a range of measures to manage contractors before they deploy into theatre and during operations. Key elements include the generation of a CSO policy and legal framework, including Joint Service Publication 567, incorporating policies on sponsored reserve, contractors on deployed operations (CONDO) and private military and security companies (PMSCs), MoD Standard 05/129 stating CONDO processes and requirements, and MoD Defence Condition (DEFCON) 697.

Furthermore, in 2012 the MoD adopted the US 'Synchronized Predeployment and Operational Tracker' (SPOT) system to improve tracking of UK contractors both before deployment and when deployed,[61] linking contractor personnel to the contracts they support, and thus generating easily accessible data on the contractor landscape in terms of locations, numbers, and the tasks or contracts they fulfil.[62]

Notwithstanding these improvements, the UK's CSO experience in Iraq and Afghanistan, although positive overall, suggests that the British government and the armed forces must continue to improve their CSO contracting, management and oversight capabilities. There still appear to be too few experienced acquisition personnel with an in-depth knowledge of the CSO commercial base to manage the funds and workload brought about by such significant CSO demands.

The private sector has to secure the long-term sustainability and profitability of its CSO business models, which, particularly in the equipment and people-support sub-markets, must be considered as part of wider service sectors. Companies need skilled, well-trained and equipped, mentally and physically prepared, and reasonably paid

61 Office of the US Deputy Assistant Secretary of Defense (Program Support), 'Synchronized Predeployment and Operational Tracker – Enterprise Suite', information sheet, <http://www.acq.osd.mil/log/PS/SPOT/Definition_SPOT_ES_InfoSheet.pdf>, accessed 10 October 2013; see Office of the US Deputy Assistant Secretary of Defense (Program Support), 'US/UK Force Generation Analysis', p. 7.

62 Commission on Wartime Contracting in Iraq and Afghanistan, 'At What Risk?', Second Interim Report to Congress, 2011, pp. 1–5.

employees to provide high-quality services.[63] They also need realistic risk assessments, grounded in an in-depth understanding of the highly complex operating environment, and a balanced perception of commercial, reputational and personal risks. Ultimately, these companies have to deliver high-quality products and services which can be sustained in theatre at acceptable costs.

The United States

The US military has seen operations in every decade since the Second World War. It is important, therefore, to ask how the US government has viewed its defence industrial support during military operations. There is also a subtle and significant difference between industry developing defence equipment and materials over a long period of time, within a specific defence programme, and the immediate support to operations offered by industry.

Former Defense Secretary Robert Gates has noted that direct contractor support to operations has been crucial to the execution of both the Iraqi and Afghan campaigns. He anticipated that counter-insurgent operations such as these would be a staple of military practice in the future, and that industry support to operations on the front line would be an increasingly significant component of the US's order of battle. In terms of the undertaking of military operations, the US government is dependent upon industry for much of its actual operational deployed force structure and, in consequence, is viewed by government as a most significant national-security resource.

In this sense, the former US Under Secretary of Defense for Acquisition, Technology and Logistics Ashton B Carter states, 'We're simply not going to go to war without contractors. We have to build that into what we call readiness, what we call training, what we call leadership, and what we call war planning.'[64]

As a signal of the importance with which the US government views its contractors deployed on operations, in 2008 the establishment of the US Army Contracting Command was initiated, with a remit to develop an operational structure and high-level process through which to conduct an expeditionary force made up, in part, of contractors.

Furthermore, since 2010, the DoD has revised its core operational contract support (OCS) policy documents and is due to release its updated

[63] Robert M Gates, 'A Balanced Strategy: Reprogramming the Pentagon for a New Age', *Foreign Affairs* (January/February 2009).

[64] Ashton B Carter, 'Better Buying Power in Defense Spending', statement before the Commission on Wartime Contracting, Washington, DC, 28 March 2011, p. 39, <http://www.wartimecontracting.gov/docs/hearing2011-03-28_transcript.pdf>, accessed 10 October 2013.

joint doctrine, 'Operational Contract Support' (originally issued in 2008), in early 2014.[65]

Despite these improvements, the US government still faces many significant challenges in its approach to OCS. In order to fully institutionalise OCS, the US government seeks to further elevate its role within DoD culture. The US Congressional Research Service (CRS) recommends that:[66]

1. Senior leadership must maintain its focus on articulating the importance of contract support in a sustained and consistent manner
2. The Professional Military Education curriculum must fully incorporate courses on operational contract support throughout its various efforts
3. Training exercises must be expanded and incorporate contractors playing the role that they would play on the battlefield.

Further, the CRS concludes that systemic changes are required to ensure effective and efficient OCS. First and foremost, the DoD must improve the integration of contractors into its operational planning and overall strategy, in order to deliver a proactive utilisation of OCS capabilities. The ability to realise such integration is crucially dependent on the availability of reliable and appropriate data systems to build a more complete contracting operating picture. Moreover, the US government has to provide a sufficiently large and technically capable workforce to manage and oversee contracts and plan for their use.[67]

Much of this political commitment to contractors on operations, however, is derived from hard-headed fiscal considerations. Indeed, if military personnel were to fulfil those roles now undertaken by industry in direct support to operations, the upfront costs to government would be high, and for each soldier deployed another would have to be in training and a third probably on leave as part of the military force structure. This

---

[65] In 2010, the DoD updated its 'Policy and Procedures for Determining Workforce Mix', which addressed contractor personnel as part of the total force, and in 2011 released a major update to the DoD instruction 'Operational Contract Support', which established roles and responsibilities for managing operational contract support. In 2012, the DoD updated its joint planning and execution policy to include operational contract support in many non-logistical functional areas, such as intelligence, personnel and engineering, and one year later developed standards for using private security contractors.

[66] See Moshe Schwartz and Jennifer Church, 'Department of Defense's Use of Contractors to Support Military Operations: Background, Analysis, and Issues for Congress', Congressional Research Service Report, 17 May 2013, p. 16, <http://www.fas.org/sgp/crs/natsec/R43074.pdf>, accessed 15 October 2013.

[67] See *ibid.*

tends not to be the case with a contractor's staff.[68] Consequently, industry on the front line is a feature driven by the realities of budgeting, government's quest for efficiencies and the inevitable dynamic of the investment appraisal.

Germany

Although there have certainly been problems in relation to this still new segment of the government–industry relationship, particularly in the initial phase of the Bundeswehr's Afghan operation, the overall experience of the German government with industry's contribution to Bundeswehr operations is highly positive. Like in the UK and the US, industry has generally acted quickly and professionally regarding the provision of UORs and much-needed material. According to government officials, problems that occurred could usually be blamed on inefficient procurement and licensing procedures, and only in rare cases on the private sector's failure to deliver against contracts.[69]

In general, the Bundeswehr operations in Afghanistan, the Balkans, the Democratic Republic of the Congo and elsewhere have highlighted to the German government the need for a more strategic, coherent defence industrial policy that considers the German defence industry as part of the overall military effort. However, the Bundeswehr's doctrine, structures and processes do not yet fully reflect this understanding.

This is particularly apparent in terms of the integration of CSO into the Bundeswehr's planning and operational conduct, which is still underdeveloped compared to equivalent approaches in the UK and the US. The Bundeswehr so far lacks a comprehensive CSO doctrine. However, various Bundeswehr doctrinal documents, including the 'TK Logistik', 'TK Unterbringung im Einsatz' and 'TK Betriebstoffversorgung', address a number of aspects which are relevant to CSO. In addition, in 2008 the commanding general of army logistics, Kurt Helmut Schiebold, presented the position paper 'Aspects of Civil Logistical Support for the Army's Operational Logistics', which describes the general challenges associated with CSO in the logistical domain.[70]

Moreover, in 2004 the working group on logistics 'Bundeswehr and Industry' generated the 'Guidelines for the Co-operation of the German Industry with the Bundeswehr in the Context of Operations and Exercises

---

[68] See Congressional Budget Office, 'Contractors' Support of US Operations in Iraq', August 2008, pp. 16–17, <http://www.cbo.gov/sites/default/files/cbofiles/ftpdocs/96xx/doc9688/08-12-iraqcontractors.pdf>, accessed 15 October 2013.

[69] Author interview with senior government officials.

[70] Kurt Helmut Schiebold, 'Aspekte ziviler logistischer Unterstützung für die Einsatzlogistik des Heeres', position paper, Bundeswehr, May 2008.

outside the Federal Republic of Germany'.[71] Its goal is to provide contractors with fundamental information about the specific operating framework in theatre. Topics covered by the guidelines include working hours, responsibility for transport, catering and equipment, as well as a checklist for the deployment of employees in crisis regions. However, as the document is not legally binding, it does not represent a substitute for a comprehensive CSO doctrine.

The participation of contractors in the Bundeswehr's planning structures and procedures prior to an operation is equally underdeveloped. The Bundeswehr has not yet established long-term, institutionalised co-operation between industry and the Joint Operations Command (Einsatzführungskommando), which conducts all operational planning in the Bundeswehr. A German equivalent of the British CONLOG contract or the US Logistics Civil Augmentation Program (LOGCAP) currently does not exist.

Companies providing support to the Bundeswehr in theatre can be categorised in two groups.[72] The first group consists of companies contracted for logistical support, including Supreme Foodservice and Ecolog. Both companies have extensive experience serving the Bundeswehr as a customer. In Afghanistan, they hold the largest-value contracts among all Bundeswehr contractors.

The second group consists of manufacturing companies, which produce the Bundeswehr's equipment, including Krauss-Maffei Wegmann (KMW) and Rheinmetall. The latter, for example, maintains the Bundeswehr's Heron UAV in Afghanistan, in co-operation with Israel Aerospace Industries. Several members of KMW's customer-service team have also provided local support to the German army since the beginning of the ISAF operation in Afghanistan.

Furthermore, in 2004 KMW and ten medium-sized German companies concluded a service co-operation agreement and created the 'Service Industry Group in Deployment Areas' (Industriegruppe Service im Einsatz – IGS E). The agreement encompasses the provision of services such as repair and maintenance work on equipment located in areas where the German army or other NATO partners are deployed.

In general, industry is viewed by the German government as a crucial partner to the Bundeswehr on operations. However, so far, this key role is inadequately reflected by the Bundeswehr's CSO doctrine and its

[71] Arbeitskreis Logistik Bundeswehr und Wirtschaft, 'Leitfaden für die Kooperation der deutschen Industrie und der gewerblichen Wirtschaft mit der Bundeswehr im Zusammenhang mit Einsätzen und Übungen außerhalb der Bundesrepublik Deutschland', 22 April 2004.

[72] See Jan Stöber, *Battlefield Contracting – Die USA, Großbritannien, Frankreich und Deutschland im Vergliech* (Wiesdbade: Springer US, 2012), pp. 129–58.

contractual design but the Afghanistan operation has certainly served as an accelerating factor for the further development of the German CSO approach.

Summary

Both UORs and CSO are key components of the British, US and German national defence efforts. The three countries share the experience that, overall, industry has responded efficiently to operational demands and has proven itself a reliable partner in theatre. With the redeployment from Afghanistan already having started, the UK, the US and Germany all face the challenge of identifying the UORs that need to be transferred into the core equipment portfolio and of managing this process effectively. Further, they must advance their UOR procurement and in-service procedures based on lessons learned.

The advancement of CSO policy is a common challenge for the British, US and German governments, too. As shown above, the US has made substantial progress in this regard – partly bringing formerly outsourced functions back 'in house' – with the UK and Germany, in particular, arguably lagging behind, not least due to the type, scope and frequency of their deployed military operations in the past.

Conclusion

This chapter has clearly shown that the British, US and German governments have substantially increased their reliance on the private sector for the provision of products and services, as well as much training, in the defence sector over the last decade. Industry has thereby become an inherent component of the British, US and German national defence efforts.

Over this period, government bodies and the armed forces have learned many lessons from engaging with the private sector in defence. These have had far-reaching implications for the government–industry relationship and the government's role as a customer. The evolution of this role has not always proceeded in a strategic manner, having often been built on ad-hoc decisions – frequently taken in response to financial pressures or operational demands. Governments face the common challenge of further advancing their customer roles according to the changing government–industry relationship in the defence sector and the evolving international strategic environment.

At the core of this reside governmental endeavours to sustain a healthy relationship with industry and to become more intelligent and effective customers in the defence market. The latter task encompasses the reform of defence procurement structures and processes, a clear identification of vital defence industrial sectors, the introduction of more objective and reliable performance indicators for their defence management efforts, the provision of a higher degree of transparency in acquisition plans and intentions, and the advancement of CSO approaches.

In addressing these issues, the British, US and German governments must generate their policies against the background of the specific national lessons learned in the defence market and the respective state of their defence industrial bases, within the framework of the existing legislation, political system and organisational culture. Therefore, although they often face similar challenges, their individual national responses differ.

As for the reform of defence acquisition, all three governments have a clearly stated preference for competition in order to ensure best value for money. Equally, the UK's defence White Paper, the US's Better Buying Power 2.0 Initiative and Germany's revised Customer Product Management system all require national procurement bodies to address the possibilities of off-the-shelf procurement and multinational collaborations, in order to cost-efficiently close capability gaps. However, instead of opting for full and open competition, procurement authorities may regularly limit competition, due to the technological complexity of many procurement projects and the nature of the defence market, with limited suppliers (sometimes only one) for certain products.

Furthermore, the US example has shown that full competition can have adverse effects on 'value-for-money' in defence and on the sustainment of a capable and flexible defence industrial base, leading to artificially low bids by companies to win the contract, and potentially causing losing companies to leave the sector. Therefore, limited competition and single-source contracts are a constant element of the British, US and German approaches to defence procurement. The latter in particular often mirror governments' interest in ring-fencing national defence industrial capabilities in order to secure jobs and sustain core capabilities, the need to classify procurement projects for national-security reasons as well as governments' efforts to control the transfer of sensible technology.

Seeking to better match this key feature of the defence market, an adequate distribution of commercial risk in such contracts and a sufficient ability to compare financial arrangements to competed tenders, governments have revised, or are in the process of updating, their single-source contracting regulations. While the UK is to publish its update of the 'Yellow Book' that lays out the government's single-source regulations for defence procurement in the coming months, the German government has addressed single-source contracts in revising the legal framework for Bundeswehr contracts, though it remains to be seen how single-source contracts are conducted by the revised CPM processes.

Beyond the awarding of contracts, the UK's and Germany's defence-procurement structures are undergoing substantial reforms. In Germany the newly created defence procurement body – Das Bundesamt für

Ausrüstung, Informationstechnik und Nutzung der Bundeswehr (BAAINBw) – has for the first time introduced military staff into the formerly purely civilian defence-acquisition structures in order to ensure appropriate consideration of the user perspective; in the UK the government has launched a year-long assessment of the reform of its DE&S organisation. After this assessment phase, it will be decided whether to transform DE&S into a GOCO entity or to leave it within the public sector as 'DE&S+'. These examples highlight that both of these governments, like the US, have understood the need to improve defence-procurement procedures in their endeavours to become more intelligent and effective customers in defence – though potentially through different steps.

The ability to adequately account for a project's through-life success is an important element in this regard. It requires governmental acknowledgement of through-life considerations for both the total costs of operating a system over its lifecycle, and its actual usefulness, reliability and availability once it comes into service. To achieve this, an effective departmental through-life project controlling system is required. Arguably, neither the UK, US nor Germany has so far been able to fully establish such a system, although improvements have been made, as in the case of the revised German CPM, which requires planners to state the estimated whole-life cycle costs for a proposed procurement project.

Moreover, independent scrutiny bodies, such as the British NAO, the US GAO and the German Bundesrechnungshof, are key stakeholders in measuring their governments' success as customers in defence. Whereas NAO and GAO reports are usually publicly accessible, and discussed, most reports by the Bundesrechnungshof remain unnoticed by the wider German public. Yet, while available reports support a better understanding of industrial capability to deliver on time and to budget, they usually do not present sufficient information – or at times any information at all – on a product's operational success, thereby limiting their added value to governments' efforts to be better informed, and thereby more intelligent, customers.

Government's ability to act as an intelligent and effective customer in defence is especially important when it opts to deploy armed forces on operations abroad. All three governments have created procurement mechanisms to address UORs. The overall experience in the UK, the US and Germany suggests that industry has responded flexibly to these demands and proved to be a reliable partner to government.

Furthermore, the advancement of CSO policy is perceived as a common challenge by the British, US and German governments alike, though their respective starting positions differ due to the type, scope and frequency of past deployed military operations. The US has relied significantly on contractors for its operations in Iraq and Afghanistan.

Supported by the independent Commission on Wartime Contracting in Iraq and Afghanistan, whose findings eventually led to the re-integration of formerly outsourced functions like the management of contractors, the US has reached a high degree of maturity in its CSO policy.

While the UK has also made substantial progress in its CSO approach, with the introduction of new policies and systems to monitor contractor activities, the need for more experienced acquisition personnel with profound CSO knowledge and management skills is still apparent. This assessment also holds true for Germany. Although the latter's reliance on contractors has been lower than in the UK and the US, the private sector must still be considered as a crucial partner to the Bundeswehr on operations. However, so far, this key role is not adequately reflected by German doctrine, structures and processes. With a more significant contractor role in future German operations being likely, German CSO policy needs to catch up with its US and UK equivalents.

Finally, this chapter has highlighted that for governments to ensure a healthy government–industry relationship in defence it will be essential to transparently communicate their intentions and plans as customers in the defence market. Companies require a certain degree of predictability to sensibly decide on their future position and commercial viability in the defence sector. In particular, governments must clearly communicate those areas they regard as core defence industrial capabilities that must be sustained on-shore and those functions they consider as 'inherently governmental', and so are unsuitable for handling by the private sector.

The US government has been very active in creating a multilayered structure of conceptual policy documents to inform the public and industry about its intentions and plans. Moreover, it has more or less clearly communicated its perception of core defence industrial capabilities and defined in detail the term 'inherently governmental functions'. As a customer in defence with a mature ability to think strategically in terms of its security and defence industrial policy, the US government has acknowledged industry's need for such information and clarity, especially as the degree of consolidation of the US defence industry is significantly higher than that in the UK and Germany.

In the UK, the government has recently recognised the relevance of the term 'inherently governmental function' to the wider debate on defence acquisition in its June 2013 White Paper on better defence acquisition.[73] However, beyond the introduction of the term in the

---

[73] Ministry of Defence, *Better Defence Acquisition: Improving How We Procure and Support Defence Equipment*, Cm 8626 (London: The Stationery Office, June 2013), para. 38, <https://www.gov.uk/government/uploads/system/uploads/attachment_data/file/206032/20130610_WP_Better_Def_Acquisition_screen_final.pdf>, accessed 10 October 2013.

acquisition discourse, it has done little to explain what this actually means for concrete outsourcing decisions. Furthermore, the current UK government has not conceptually determined – in the form of a policy paper – what it considers to be core defence industrial areas. However, existing contractual arrangements, such as Team Complex Weapons and contracts placed on-shore for nuclear reaction and submarine work, give an implicit indication of such areas.

Germany, arguably, did not have an organisational culture in defence in the past that continuously supported a clear and transparent communication of policy plans and intentions. Neither does a prioritised list of core defence industrial capabilities, nor a single, precise definition of 'inherently governmental functions' yet exist. However, the German government seems to have acknowledged this strategy gap and is in the process of addressing it. Work on a German defence industrial strategy is also underway in the BMVg and the Bundeswehr's new Integrated Planning Process may offer an opportunity more realistically to prioritise core national defence industrial capabilities.

In conclusion, it seems reasonable to state that British, US and German governments alike acknowledge the imperative to constantly advance their roles as customers in the defence market to ensure good value-for-money for the taxpayer and the provision of high-quality equipment and services in support of the national defence effort. Their national policies in the areas of 'customership' are more or less coherent, with only minor differences reflecting their respective national contexts.

Accordingly, governments' customer roles do not appear seriously to hinder the ability of defence enterprises to operate effectively and efficiently. Neither do they generate fundamental friction in defence collaborative and co-operative projects, although it has been shown that national defence industrial ring-fencing – for example, through the placement of on-shore single-source contracts – is a somewhat limiting factor in this regard.

Industry is naturally attracted by the US as a customer, with the country having developed a mature conceptual security and defence industrial policy framework. The US government is, therefore, able to provide industry with a sound understanding of its intentions and plans, as well as the size of the US defence equipment budget. However, the chapter has also shown that the British and German governments are working to address the shortfalls in their customer roles to sustain a healthy government–industry relationship in defence, thereby encouraging companies to remain and further invest in European defence markets.

III. THE STATE–DEFENCE INDUSTRIAL RELATIONSHIP: GOVERNMENT AS SPONSOR

The previous chapter considered the notion of the modern state as the customer of defence industries, typically generated through some kind of commercial 'customer–supplier' relationship. This chapter conceptualises the state as the sponsor or 'enabler' of the very businesses from which it receives defence equipment, material and services.[1] This difference in emphasis hints at the interdependence of state and defence businesses beyond the conventional dynamics of the marketplace, reinforcing the view offered in Chapters I and II of the state–defence industry relationship as complicated and subject to competing factors.[2] Indeed, the sense that government finds it desirable, or even necessary, to sponsor this element of its industrial base suggests that the relationship exists beyond just the commercial or managerial spheres. To what degree this is the case, and for what purpose, is explored in this chapter.

It is important to remember that the concern, here, is with the defence industrial base operating from within the state's borders, the companies that manufacture equipment and consumables, and those that provide services to the military. The concept of a 'sponsor' is, of course, different from that of a 'customer'. It is, however, the very nature of the defence market that causes analysts and decision-makers to reach beyond the 'customer' label when attempting to understand the relationship between government and industry. The sense of just one customer (the government), limited suppliers and limited information precludes the dynamics of a perfect market from working, and the lethal nature of

[1] By 'sponsor' the authors mean simply government's commitment, or otherwise, to support and promote defence industries through the generation of tailored policies, processes and activities.

[2] See James F Nagle, *A History of Government Contracting* (Washington, DC: George Washington University Press, 1999).

the business of defence keeps it apart from other markets, even those in which government could be a major player.[3]

Government sponsorship of defence industries is therefore concerned with the functions of securing, promoting and enabling private-sector organisations and, through them, the competencies and skills to support the military through the provision of goods, services, materials and know-how.[4] Of course, it may also be concerned with ensuring a competitive edge against international competitor firms. Such sponsorship is enabled through government funding and state-level support to defence industrial activities and, consequently, the businesses in question have greater confidence to raise funds and invest. Moreover, and significantly, financial markets and other stakeholders are assured that these companies are healthy and viable for the long term as a critical component of national defence – *for* the state, but not *of* the state.

This chapter explores the conditions for government sponsorship, as well as the mechanisms through which that sponsorship is enacted by government. The implications of this practice on both the state and businesses are then considered, as is the question of who sponsors whom.

National Variations in Sponsorship

The conditions for sponsorship across defence vary, but much of the research literature points to a number of common features and considerations.[5] The first of these is the recognition that the relationship between government and defence and national-security companies is transactional. Consider the example of the United States where, as is well known, defence is big business. The Department of Defense (DoD) contracts for over $450-billion-worth of goods and services each year, processing

[3] For a historical perspective on this point, see Merton J Peck and Frederic M Scherer, *The Weapons Acquisition Process: An Economic Analysis* (Cambridge, MA: Harvard University Press, 1962). See also Eugene Gholz, Harvey Sapolsky and Caitlin Talmadge, *U.S. Defence Politics: The Origins of Security Policy* (New York, NY: Routledge, 2008).

[4] See Terrence R Guay, *Globalization and Its Implications the Defence Industrial Base* (Carlisle, PA: Strategic Studies Institute, US Army War College, February 2007).

[5] For a general discussion of the complexities of defence and security questions in periods of government austerity, see Michael Codner and Michael Clarke (eds), *A Question of Security: The British Defence Review in an Age of Austerity* (London: I B Tauris, 2011). See also Mark Abramson and Roland Harris, *The Procurement Revolution* (Lanham, MD: Rowman and Littlefield, 2003). A good contextual comparative volume is Ethan B Kapstein, *The Political Economy of National Security: A Global Perspective* (Columbia, SC: University of South Carolina Press, 1991).

in excess of 14 million commercial invoices per annum.[6] The general defence ledger is comprised of close to 60 million transactions every year, indicating that the government's relationship with its commercial suppliers for defence is substantial, highly systematic and institutional.[7] This established financial and accounting infrastructure is one of the key ingredients in – and a prime condition for – this type of sponsorship, as are the culture, commitment and mechanisms required to handle such a volume of transactions.[8]

It is significant that the US Federal Acquisition Regulation (FAR) mandates the manner in which these transactions are conducted and supervised so that any company wishing to do business with the DoD must comply with these detailed regulations. However, economists would, of course, recognise that even perfect markets have formal and informal rules which frame the engagement of buyers and sellers. Although significant, this transactional infrastructure – bespoke to the defence sector – is therefore a function or practice of state sponsorship, rather than a reason for its existence. Instead, there are a number of distinct drivers of government sponsorship of the defence industrial base, beyond the transactional, that need to be considered.

In Germany, significantly, two very distinct categories of defence industrial sponsorship can be distinguished (see Figure 3 on the following page): the non-defence-specific policies that aim to provide a beneficial political, economic and societal climate for business, and defence-specific policies that address the particular demands of the defence industrial base in order to promote a commercially viable sector. Each of these categories encompasses a large range of policies, which either directly or indirectly affect the German defence industrial base and its relationship with the state.

Moreover, it is important to stress that many drivers of the development of the national defence industrial base, both in Germany and the UK, reside beyond the sphere of direct governmental influence, either because legal responsibility rests with other stakeholders, who may be averse or even hostile to government influence, or because they present highly complex social processes, which cannot be influenced through government policies. However, if applied coherently, the combination of policies from both categories of a defence-industrial-sponsorship cluster presents governments with a

---

6 GAO, *Defense Acquisitions: Assessments of Selected Weapon Programs*, GAO-07-406SP (Washington, DC: GAO, March 2007).

7 See Department of Defense, 'Quadrennial Defense Review Report', February 2006.

8 See GAO, *Defense Acquisitions: Assessments of Selected Weapon Programs*, GAO-08-467SP (Washington, DC: GAO, March 2008).

Figure 3: Elements of Defence Industrial Sponsorship.

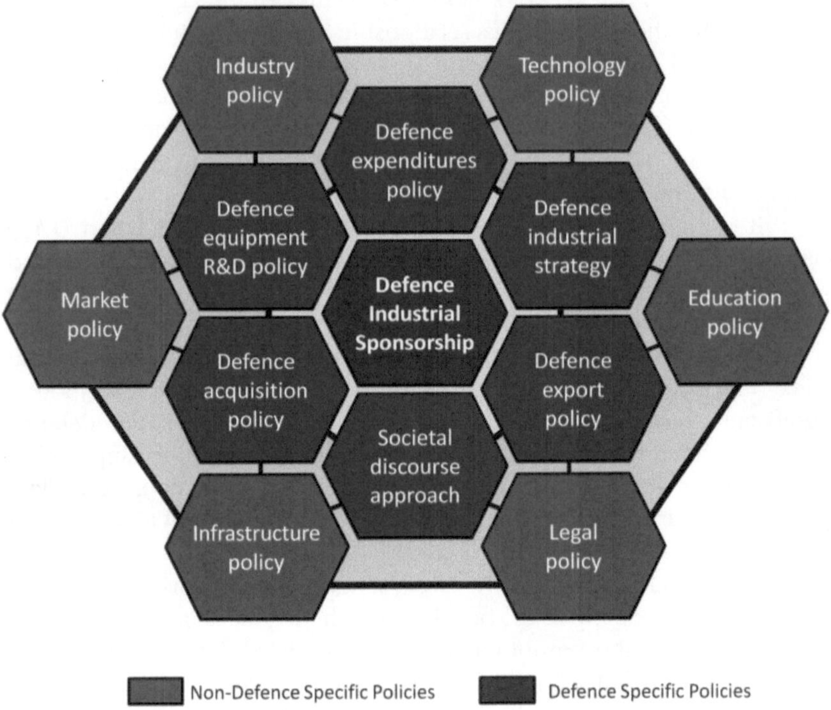

Source: The authors, 2013.

powerful instrument to support a functioning defence market, a healthy defence industry within its borders and, ultimately, a sustained national defence effort.

This is a neatly tessellating picture of defence, suggesting that all ingredients fit into a coherent, logical whole. The reality is more complicated. This complexity, the interdependency between the state and defence industries, and the sheer volume of transactions are themes that consistently reappear within the modes of government sponsorship – including research-and-development (R&D) practices, acquisition policies, import and export policies, and defence industrial strategies – as discussed below. As such, it is more like a multidimensional, multidisciplinary jigsaw where the pieces do not naturally fit together, and there is no guiding picture to assist with assembly. Governments must therefore respond in the best way they can, with imperatives varying between states. The UK's response, for example, is briefly considered in Text Box 5.

Text Box 5: What Makes a Firm 'British'?

The challenges of delineating the defence industrial sector were explored in the Introduction to this Whitehall Paper, but here it is necessary to explore the elements required for an enterprise to be considered 'British'.

There are several different forms of enterprise that are relevant. There are privately owned businesses based in the UK, such as Marshall of Cambridge and Martin-Baker. There are multinational joint ventures, of which MBDA is the most important, and there are public companies, headquartered in the UK, quoted on the London Stock Exchange, and owned in large part by British nationals and institutions. However, such firms can also have major foreign stockholders and, in 2002, the government lifted the limit on the foreign-share ownership of Rolls-Royce and BAE Systems, although the stipulation remained that no single foreign shareholder could hold more than 15 per cent of the company. In principle and practice, shares in defence firms can be held by foreign sovereign-wealth funds, although of course the usual national rules apply regarding the need to identify any shareholders who have more than a specified percentage of the total. Finally, there are the UK subsidiaries of foreign companies. The latter were a negligible consideration before the end of the Cold War, but since then there has been extensive investment by European and American firms in UK defence-oriented facilities. While BAE Systems remains the MoD's largest supplier by far, the next largest defence employers in the UK are Finmeccanica and Thales.

In the 2002 Defence Industrial Policy document, the government both encouraged further foreign defence investment in the UK and acknowledged reality by announcing that any firm which added significant value to industry would be treated by the government as a British company. One of its key conclusions was that: 'The UK defence industry embraces all defence suppliers that create value, employment, technology or intellectual assets in the UK. This includes both UK- and foreign-owned companies.'

As defence and national security is, in part, a highly technological phenomenon, government's role in securing the research and development of technologies applicable to military use is highly significant. So, too, are national preferences in defence procurement, regarding, for example, the companies selected by governments for contracts or the commercial capabilities they champion. Notions of industrial participation in the state – policies relating to taxation and state investment – can also reveal certain elements of government sponsorship. Of significance, too, are modern concepts of partnering between government and businesses as a function of sponsorship. In these instances, government shows a specific preference for a particular company (possibly post-competition), awarding it a long-term partnering contract for the provision of equipment or services, usually on an availability or output basis. Moreover, analysts often consider government support for exports as an overt example of government preference for, or sponsorship of, on-shore businesses in contrast to those registered in other countries. These are all important issues, addressed in the analysis below under the headings of Research and Development, Defence Acquisition, Imports and Exports, and Defence Industrial Strategies.

Research and Development

Warfare is concerned with exploiting an advantage over an enemy. That advantage has often been technological. Therefore, the need for R&D to secure that technological advantage is, and always will be, a pivotal theme within defence.[9]

The Scale of US Research and Development

More than any other nation on earth, the US is wedded historically to the reputational, operational and power-discourse benefits of military R&D. Consequently, R&D investment by the federal government is a major pillar of the state's sponsorship of the US industrial base. In fact, for most of the twentieth century – throughout the Second World War, the Cold War, the asymmetrical War on Terror and beyond – and to the present day, the US's national-security and defence strategy has been built around the idea of developing, and thereafter sustaining, substantial technological superiority over potential adversaries and allies alike. For example, during the Cold War, successive US administrations took conscious decisions to offset the Soviet Union's supposed superiority in military manpower by investing in technology.[10] Even today, China, as an emerging geopolitical rival to the US, has 2,285,000 trained military men and women under arms, whilst the United States has forces of 1,563,996, according to data from 2010. The US, however, spends upwards of $712.8 billion on its military components compared to China's $98.36 billion, indicating that the US remains wedded to a highly technological military, underscored by effective investment in R&D.[11] Consequently, US policy is connected overtly to the development of advanced technologies, requiring the government to have a sense of how these technologies could, or should, be used in the future and by whom. US research and applied investigations are therefore both highly technical and scientific, but also deeply embedded within the social sciences – as well as even more modern academic disciplines, such as business studies and new media – as horizon-scanning and concepts of

[9] See National Research Council of the National Academies, *Science and Security in a Post-9/11 World* (Washington, DC: National Academies Press, 2007).

[10] See Jacques S Gansler, *Democracy's Arsenal: Creating a Twenty-First-Century Defense Industry* (Cambridge, MA: MIT Press, 2011), p. 245. See also Paul Bracken, *Technological Innovation in National Security* (Philadelphia, PA: Foreign Policy Research Institute, 2008).

[11] See International Institute for Strategic Studies, *The Military Balance 2011* (London: Routledge, 2011).

future operating environments become increasingly significant. Simply put, as technology and its use spreads globally and rapidly, the US is keen to stay ahead, offensively and defensively, of this transformational trend, in order to ensure its security and that of its markets and suppliers. This underscores the extent to which the US government invests in R&D in the private sector – the cornerstone of the American state's sponsorship of its defence industry.

The scale of this investment is truly staggering. Research, development, testing and evaluation task-order agreements in the US equate to about $70 billion a year – about a tenth of the annual defence budget.[12] This is about the same size as China's entire defence budget, and $10 billion more than the UK's budget. Indeed, research and development for the maintenance of technological advantage in the US is a serious business, driven, overtly, by governmental paternalism towards its enabling industrial base.

Since the end of the Second World War, industry in the US has undertaken about 70 per cent of the R&D funded by the DoD, while laboratories funded directly by the government have carried out about 25 per cent and US universities about 5 per cent.[13] As such, on the basis of the raw data at least, government directly sponsors industry, usually defence businesses, to undertake the overwhelming majority of its research. Moreover, as discussed in the previous section, unlike the commercial sector, the defence market is characterised by a single buyer – government – and a small number of entrenched suppliers in each critical area of capability. Indeed, given its objective to secure and maintain technological superiority in defence, the government must take on a uniquely proactive, shaping role. In its sponsorship of defence R&D, the government recognises the fundamental point that most businesses cannot afford to spend billions on R&D from their own shareholders' funds in the hope that government may wish, in the future, to buy its products. The risks of failure are simply too great for companies to invest in this manner. Consequently, to engage industry in the research, development and production of highly technical, high-risk weapons systems, government has to sponsor both the initial research, the applicability of technological development and – sometimes post-competition – a high-risk development programme encompassing manufacture, military use, through-life support arrangements and eventual disposal. This starts, however, with

---

[12] Based on the United States' national defence budget estimates for the fiscal year 2006–07, Office of the Under Secretary of Defense (Comptroller), April 2005. See Gansler, *Democracy's Arsenal*, p. 254.

[13] See National Research Council of the National Academies, *Beyond Fortress America: National Security Controls on Science and Technology in a Globalized World* (Washington, DC: National Academies Press, 2009).

government's commitment to the research expenditure that companies operating under financial constraints and treasury parameters cannot bear themselves. Nowhere is this better recognised than in the US.

In the US, there is a correlation and interdependence between a strong economy, federal expenditure on national security, investment in defence R&D, and defence technological spin-offs into the commercial world and non-defence markets. For example, the Electronic Numerical Integrator and Computer (ENIAC) was funded by the US Army as far back as 1945, and was one of the world's first electronic digital computers. Since then, the DoD has been a major sponsor and funder of technological advances in the computer industry. In parallel, it has funded significant advances in software development and manufacture, to the extent that well over 50 per cent of the entire development costs in relation to the Joint Strike Fighter (JSF) are associated with software design and development. The Defense Advanced Research Projects Agency (DARPA) is credited with developing its own virtual network from 1974 onwards, which contributed significantly to the development of the Internet. The development of the global positioning system (GPS), semiconductors, jet engines, coolants, nuclear power, freeze-dried foods and standardised shipping containers can also be highlighted as examples of major contributions by US defence R&D to civil commercial innovations and products during the recent period of high modernity, all of which have helped to drive not just the US economy but global growth and development as well.

Yet, despite the broader economic benefits, the US government has to sponsor defence and national-security R&D because the market rationale mitigates against the volume, value, depth and breadth of investment required from commercial businesses. Government has to step in if the US is to deliver its core policy of maintaining its technological advantage in defence. Consequently, there are, in essence, three core objectives for DoD investment in R&D if this strategic outcome is to be met. First, the DoD must stimulate innovation. Second, government must avoid surprises. Third, and as a culmination of the successful accomplishment of the preceding two objectives, the US must stay ahead of others.

The US has a good record of stimulating technological innovation, paradoxically, as both the champion of free-market economics and as a government funder – and possibly also champion – of research. The Small Business Innovation Development Act of 1982 has grown from a small-and-medium-enterprise (SME) development project of $5 million per annum to a federal government programme of $2 billion, with over half of this going to small businesses in the US defence sector. The DoD also sponsors over 35 per cent of basic research in computer science and maths at American universities, as well as over 30 per cent of the basic engineering research conducted at graduate and postgraduate level.[14] Government, therefore, serves as a sponsor of both educational research activities and those undertaken across smaller, niche businesses. Many of these companies and research activities form part of the value chains of the large defence industrial prime contractors and integrators in the US. By investing to stimulate innovation and technological problem-solving, the US government has been able to partly match complex and long defence-procurement cycles to relatively shorter modern technology-development cycles, driving, it is hoped, the retention of US technological hegemony.

DARPA was established within the DoD in 1958 to serve as the programmatic focal point for R&D across national security in the US, to assure the lead in the application of high-end technology for military and defence purposes, and to actively interdict technological surprise from identified and potential future adversaries. The agency, today, runs no laboratories of its own, but funds programmes in industry and at universities with a budget of approximately $3 billion per annum.[15] It supplements federal R&D programmes embedded within the single services and works across the spectrum of research, from horizon-scanning and initial science to the generation of integrated technological solutions. Interestingly, and significantly, DARPA does not base its work on responding to defined user requirements, since few possess the skill to think about requiring things that they have not previously seen, experienced or conceptualised. Instead, by focusing on the 'what if' rather than the 'what' of a defence requirement, the agency seeks to avoid military and technological surprises.

It is also important to American politicians and military leaders that the US stays ahead of its known and potential rivals (and, indeed, even its friends). Through its massive and sustained investment in R&D, in the period after the Second World War the US was able to supersede other countries' technological ambitions. In fact, throughout the Cold War, the US invested more in technological development than the whole of Europe combined.[16] Today, of course, the US still enjoys a significant budgetary advantage.

However, a dominant feature of a globalised world, in contrast to the bipolar international system associated with the Cold War, is the levelling of technological leadership and the mid-to-long-term elimination of the US's comparative advantage.[17] Moreover, highly portable

[14] See David Mowery, *Military R&D and Innovation* (Berkeley, CA: University of California Press, 2008), p. 19.

[15] See Department of Defense, 'Annual Industrial Capabilities Report to Congress', March 2008.

[16] See Gansler, *Democracy's Arsenal*, p. 268.

international sources of finance, as well as the mobility of skilled labour, will test the US's dominance in the short-to-mid term as well, to the extent that DARPA believes that staying ahead, and relevant, is the US's greatest challenge and one that is, in part, immune to the levels of investment put forth. Consequently, the DoD is developing processes and protocols to enable much more advanced tracking and harvesting of commercial and non-US R&D, through long-term partnering agreements and alliances, which will be discussed later in this chapter. The direction of travel for R&D, on the tide of globalisation, is towards interdependence rather than independence. The question of how a country can stay ahead, despite massive investment, when technology is portable, information is transferable and systems are interdependent has still to be grasped by policy-makers in the US.

It is clear from the discussion so far that the funding of R&D for defence and national security is a longstanding policy commitment of the US and a key component of the government's ongoing sponsorship of its defence and national-security industrial base. There are other elements, of course, which are addressed below, but the significance of the commitment of the US to the stewardship of R&D should not be understated. Recognising the centrality of this topic to the notion of government sponsorship of defence industry, the question of how the UK and Germany measure up in relation to the US must also be addressed.

The UK's Humble Research and Development for Defence

The UK government has a longstanding tradition of assisting its industry by funding research activity, the results of which find their way back to industry and help to provide the intellectual foundations for new projects. The distinction between research and development can be elusive, but, in essence, research spending concerns projects not yet associated with any particular system, often involving, scientifically, work at the level of the individual material or component, while development, by contrast, relates to a scientific or technological contribution to a specific programme or project.

The UK has not seen defence research as a covert means of providing support to industry in general; indeed, it has long sought to focus defence money on areas of specific interest to defence but of little importance beyond that community. For reasons of financial self-interest, the Ministry of Defence (MoD) has always wanted the private sector to spend more of its own funds on research, while also recognising that it may not be easy for a board to justify spending that may not generate a return on a product for fifteen years or more, if ever. There is only one UK customer for defence equipment, and if that customer is not interested in a particular offering, the chances of overseas sales are slim. Thus the MoD

Table 6: MoD Spending on Research (£ million, 2013 prices).

	2004–05	% of Defence	2009–10	% of Defence
Defence spending outturn	32,515	100.00	40,246	100.00
Net expenditure on research	524	1.61	575	1.43
Of which spent inside MoD	141	0.43	166	0.41
Of which spent extramurally	383	1.18	409	1.02

Source: DASA, *Defence Statistics 2011.*

has recognised the need to spend its own money in areas where significant private-sector spending is likely to be limited.

Research is not a field of defence which has fared well as resource pressures have intensified, having been awarded a steadily declining share of the defence budget. Table 6 presents data taken from Tables 1.3 and 1.7 of *Defence Statistics 2011,* produced by the MoD's own Defence Analytical Services Agency (DASA), covering the period in which the UK was using what it terms 'Resource Accounting and Budgeting' – the governmental equivalent of commercial accruals accounting.[18] This involved an explicit distinction between capital spending and spending on other resources (such as labour and consumables), whose value was used quickly.

Table 7 is taken from Table 1.5 of DASA's *Defence Statistics 2011* and shows that the MoD's Defence Science and Technology Laboratory (DSTL) was not involved with capital expenditure.

Table 7: Spending Allocation by the Research Top-Level Budget, 2004–10.

	2004–05		2009–10	
	Total (£ million)	Science, Innovation Technology (£ million and % of defence)	Total (£ million)	Science, Innovation Technology (£ million and % of defence)
Request for Resources 1: Provision of Defence Capability	37,211	506 (1.36%)	43,429	466 (1.07%)
Resource Defence Expenditure Limits	30,860	506 (1.64%)	35,890	466 (1.30%)
Capital Defence Expenditure Limits	6,351	0 (0.00%)	7,539	0 (0.00%)

Source: DASA, *Defence Statistics 2011.*

[18] Accruals accounting was introduced in the 1990s to replace cash accounting as part of the UK's new public-management reforms. See Bill Kincaid, *Dancing with the Dinosaur: How to Do Business with MoD in the Smart Procurement World* (Newcastle-upon-Tyne: UK Defence Forum, 1999).

Reduced spending on research does not appear to have reflected any policy decision to that effect. When money is short, research is vulnerable because it is not easy to demonstrate that hard results have followed from the efforts involved, and because so many of these activities are not tied into long-term contracts.

MoD-funded research should not be seen entirely as support for industry; indeed, most of it appeared, under the Labour government (1997–2010), to be devoted to the generation of informed advice, especially helping the MoD to set intelligent requirements that were both demanding and feasible. During this time, there was an effort to analyse the different benefits and outputs of research spending, and to allocate shares to each. In 2004, the National Audit Office (NAO) published the table on the opposite page (Table 8) using information provided by the MoD. While the allocated shares appear somewhat arbitrary, even fanciful, the table is useful as a means of identifying areas in which the MoD needed (and still needs) advanced knowledge, and as an indication that, even in 2004, the MoD was of the view that only about one-fifth of its research spending was of direct benefit to industry.

A further relevant publication was the 2006 White Paper on Defence Technology Strategy (DTS) – a complement to the 2005 Defence Industrial Strategy White Paper.[19] This cited the finding that the quality of a country's military equipment reflected its level of R&D spending, and generally was more oriented towards the requirement to recognise the needs of industry in order to develop and produce leading-edge equipment. It emphasised that industry should spend more of its own money on research and, by articulating some of the priority needs of the MoD, sought to reduce the risk to industry of investment in research. The DTS did not include an estimate of the MoD funding needed for its implementation and had been generated by the then minister for defence procurement, Lord Drayson. Reportedly frustrated by the reluctance of others to move forward with his ideas, he left politics soon thereafter and the DTS was apparently forgotten.

The limited impact of the DTS did not mean that it lacked criteria for the prioritisation of research activity. The UK has three significant industrial sectors which it has recognised to be of strategic importance and through which it needs the capability to design, develop, produce, test and support capable systems, including nuclear weapons, their submarine delivery platforms and 'complex weapons' – that is, guided missiles. While each sector is handled on an ad-hoc basis, there is

[19] MoD, 'Defence Technology Strategy for the Demands of the 21st Century', 2006, <http://www.science.mod.uk/modwww/content/dts_complete.pdf>, accessed 10 October 2013.

Table 8: The 'Seven Outputs'.

Category	Output	Description	Output Owner	Share of Spending (%)
Advice	1. Hot topics	Specialist scientific advice for Ministers and key decision-makers on strategic and politically sensitive issues.	Chief Scientific Advisor	3
	2. Policy and planning	Scientific advice and analytical support to policy and decision-makers.	Policy Director	0.5
	3. Capability management	Specialist advice on equipment capability planning and management such as operational analysis.	Deputy Chief Defence Staff (Equipment Capability)	51
	4. Availability	Scientific advice across the department and armed forces in support of wider activities such as procurement.	Science & Technology Director	9
	5. Technology awareness	Technology-watch activities to ensure access to the global technology base and to interpret and communicate the consequences for defence.	Science and Technology Director	13
Technology	6. Technology in the supplier base	To ensure the availability of appropriate technology work in the UK defence supplier base to meet UK defence needs.	Deputy Chief Executive, Defence Procurement Agency	14
	7. Innovative solutions	Encouraging the exploration of innovative solutions to create new or better military staff equipment capabilities.	Deputy Chief Defence Staff (Equipment Capability)	6
	Other			3.5

NB: For the MoD, 'The intended outcome of "technology" outputs is the generation of technology, systems or solutions. The intended outcome of "advice" outputs is to inform a decision-making process'.

Source: National Audit Office, *Ministry of Defence: The Management of Science and Technology*, report by the Comptroller and Auditor General, 10 March 2004, p. 9.

recognition that all require a certain amount of research or, rather, pre-project funding. There are also signs, but as yet no confirmation, that the government is ready to support the continued ability of UK industry to make a major contribution to collaborative projects in the area of airframes and aircraft propulsion systems. Finally, the UK recognises that it must fund research in some important but niche areas, in particular chemical- and biological-weapons defences and, increasingly, cyber-capabilities.

The MoD has long had a policy commitment to the promotion of SMEs in defence as a source of innovation and agility, although there is no legislated requirement to place any specific number or share of contracts with them. Since around 2005, there has also been a demand for new solutions to meet the operational challenges in Iraq and then Afghanistan. An initiative begun under the Labour government, and continued under its successor, was the Centre for Defence Enterprise (CDE) scheme. The CDE was set up in 2008 to allocate small amounts of MoD funding (a large grant under this scheme would be £1 million) in response to research proposals, especially from SMEs. The distinguishing characteristics of the CDE are that the application process is straightforward and that it promises a quick decision in response. It is intended to 'encourage the rapid delivery of cutting-edge research and development in support of front line operations. Since then it has provided funding of more than £23.5m – 43 per cent of which has gone to SMEs.'[20] The CDE recently placed its 500th contract with ITSUS, a small Welsh company which will receive funding of £104,000 to look at ways of improving existing IT network systems for military operations.

The CDE also offers marketing support to help companies develop their work into usable products, although there remain, in the UK defence sector, some general concerns about the 'valley of death', in which research funding ends but development funding is lacking, meaning that new knowledge is not exploited.

The MoD aimed to make its own research funding go further by persuading industry and others to put their own funding into projects. Clearly, this was an aspect of research funding that could not be viewed as part of the sponsorship of industry as such. The joint funding approach was most conspicuous in the Defence Technology Centres, of which there were four in 2012: Systems Engineering for Autonomous Systems, Human Factors Integration, Electro Magnetic Remote Sensing, and Data and Information Fusion.[21]

Despite the limited funding devoted to research, this is an area of the defence effort that causes significant concern to the defence industry, which is aware of how many of the products of today were

---

[20] MoD, 'Innovation on Show from the Centre for Defence Enterprise', 28 March 2012, <http://www.mod.uk/DefenceInternet/DefenceNews/EquipmentAndLogistics/InnovationOnShowFromTheCentreForDefenceEnterprise.htm>, accessed 10 October 2013; see also the CDE website, <http://www.science.mod.uk/engagement/enterprise.aspx>, accessed 10 October 2013.

[21] MoD, 'Defence Technology Centres', <http://www.science.mod.uk/engagement/dtcs.aspx>, accessed 11 October 2013.

built on research-and-technology (R&T) spending from twenty years or more in the past. There was considerable unease about the diminution of this 'seed corn', articulated by industrial associations including ADS (a trade organisation for the UK aerospace, defence, security and space industries). Recognising the validity of some of these concerns, the coalition government, in its 2012 *National Security Through Technology* White Paper, announced that the science-and-technology budget would not be allowed to fall below 1.2 per cent of the defence budget and that it intended to raise spending slightly during the period up to 2014–15. The document also recognises the value of R&T capabilities in allowing the MoD to react quickly to new hostile capabilities, implying that an important role of R&T spending is to help British industry to deliver what the MoD needs.[22]

The British government recognises that it cannot be the norm for companies themselves to finance the development of major items of defence equipment. The period of time during which returns can be made on production contracts is too long and the technical and commercial risks are too high to allow this. Hence, the MoD recognises that it must normally pay development costs as they occur, or soon thereafter, and contracts make provision for this. However, as noted in Chapter II, the British government expects companies to take on the risks of fixed- or firm-price development contracts, and is rarely ready to commit to cost-plus contracts, even with incentive arrangements for good performance, which are commonplace in the US.

The German Approach

It seems clear that equipment-related R&D expenditure has direct implications for the well-being and sustainability of a country's defence industrial base. However, German government spending on *Wehrtechnische Forschung* (defence technological research) has been in constant decline in recent years. From 2009 to 2012, this fell by 15.3 per cent from €1,097 billion to €923.2 billion. In the same period, the German government's total civilian R&D expenditure rose by 17.2 per cent, reflecting a clear political preference for civilian R&D.[23] It seems reasonable to assume that this trend will continue, given the pressure on public budgets.

---

22 See, for instance, para. viii of the Executive Summary, MoD, *National Security Through Technology: Technology, Equipment, and Support for UK Defence and Security*, Cm 8278 (London: The Stationery Office, February 2012), p. 9.

23 Bundesministerium für Bildung und Forschung (BMBF), 'Ausgaben des Bundes für Wissenschaft, Forschung und Entwicklung nach Förderbereichen und Förderschwerpunkten', <http://www.datenportal.bmbf.de/portal/Tabelle-1.1.5.html>, accessed 11 October 2013.

A German defence industrial strategy must also make reference to the German government's future approach to defence R&D. Given that the financial resources attributed to defence equipment-related R&D are declining, a forward-looking approach to defence R&D is required. Through the integration of security, defence and civil activities, allowing for the add-on exploitation of civil innovation and the dual-use of defence innovations, core defence industrial capabilities can be supported and the timely delivery of cutting-edge technological equipment and material to the battlefield can be ensured.[24]

As outlined in its departmental research plan 2013, the German Ministry of Defence's (BMVg) approach to defence-equipment R&D aims to provide the required scientific and technological knowledge and capabilities to make sensible, intelligent and cost-efficient procurement decisions; to ensure the timely identification of the threat and capability potential of emerging technologies; and to generate technologies for future and system concepts for the acquisition of defence equipment.[25] These tasks highlight the importance of defence equipment R&D for the defence industrial base. Although the BMVg's defence equipment R&D is not directly conducted by industry, it serves as an enabler and facilitator for industry's innovation systems. In this sense, corporate and public defence R&D activities form a joint base for cutting-edge technology in defence.

The German approach to defence-equipment R&D is based on three pillars: departmental research by federal institutions, funding for external research institutions, and project-related research funding for third parties.

For the second category, the Fraunhofer Group for Defense and Security is of particular importance, as reflected by its 57 per cent share of institutional research funding in the 2013 defence budget.[26] Comprising ten member institutes, the group functions as a centre of excellence for the security of infrastructure, protection of people, crisis management and

---

[24] By 'cutting-edge' the authors mean battle-winning technologies, when compared to the capabilities fielded by an adversary.

[25] See Bundesministerium der Verteidigung, 'Ressortforschungsplan für 2013 ff', 26 April 2013, pp. 5–6, <http://www.bmvg.de/resource/resource/MzEzNTM4MmUzMzMyMmUzMTM1MzMyZTM2MzIzMDMwMzAzMDMwMzAzMDY4NmE3MTcyMzY3ODZmNmEyMDIwMjAyMDIw/Ressortforschungsplan%20BMVg%202013%20ff.pdf>, accessed 15 October 2013.

[26] See Deutsche Bundesregierung, 'Entwurf eines Gesetzes über die Feststellung des Bundeshaushaltsplans für das Haushaltsjahr 2013 (Haushaltsgesetz 2013)', Bundestags Drucksache 17/10200, 10 August 2012, pp. 113–15.

surveillance, and concurrently develops cutting-edge technologies and ambitious concepts addressing both civil security and defence.[27]

Given the limited financial resources available for R&D, one of the key challenges for the German government in its approach to defence equipment R&D appears to be not to unduly restrict long-term, expensive research projects with a high potential value in the future, but with limited direct added value in favour of short-to-mid-term research projects with directly perceivable operational benefits and lower financial risks. In order to find a balance between short-, mid- and long-term research and to improve the alignment of research efforts with the current and future requirements of the German federal armed forces, the government must accept a certain level of financial and technological risk in its defence equipment research programme and clearly communicate the armed forces' future capability profile and operating environment to innovation stakeholders.

Furthermore, the German government faces the challenge of providing improved access to the civil innovation system. This requires, first and foremost, an improved awareness of emerging civil technologies and a better understanding of their add-on potential in the defence sector. Consequently, the institutionalisation of exchange processes between the BMVg, its planning, acquisition and R&D bodies, the defence industrial base and the innovation systems of the relevant commercial sector should be pursued. Conceptually, this demands a broader understanding of innovation providers in defence, an aspect that should also be addressed by a country's stated, or emergent, defence industrial strategy.

Defence Acquisition

The defence acquisition process itself must be regarded as an instrument of defence industrial sponsorship as it binds government to industry. Two elements are of particular relevance in this context: the level of efficiency in the defence acquisition process – a key subject of current defence acquisition reform – and the design of contract-awarding mechanisms, which governments can either explicitly or implicitly use to give domestic defence businesses preference in the awarding of contracts. Both elements have far-reaching implications for companies' business conduct and long-term commercial viability in defence.

Defence Acquisition and the United Kingdom

In the UK, there has been no explicit policy of this nature since the time of Peter Levene as chief of defence procurement. The MoD appears anxious

[27] Fraunhofer VVS, 'Fraunhofer Verbund Verteidigungs- und Sicherheitsforschung VVS 2012/2013', p. 2, <http://www.vvs.fraunhofer.de/fileadmin/media/vvs/Downloads/2012_VVS_Broschuere_de_online.pdf>, accessed 11 October 2013.

to stress to British industry that it will receive no special treatment. Moreover, in the face of the European Directive on Defence Procurement, the UK abandoned its Industrial Participation Policy, even for suppliers outside the European Union. Instead, the *National Security Through Technology* White Paper encouraged foreign defence firms to invest in the UK, and soon afterwards this was labelled as the government's Industrial Engagement Policy. However, in specific instances, employment and other industrial concerns have likely played an important role in UK decisions, and could well do so in future on a case-by-case basis.

The MoD no longer publishes estimates of UK defence imports, which would give some indication of any change in its readiness to consider foreign goods. There is therefore a need to fall back on partial information, such as the UN Register of Conventional Arms. Unfortunately, this covers only the period from 2007 to 2010, does not address important systems such as transport aircraft, does not address value and, in any case, contains somewhat incompatible information: Table 9 shows the contrasting US and UK reports of British defence imports from the US of the seven categories of equipment covered by the register.

Data from the Stockholm International Peace Research Institute (SIPRI) – based on the institute's own trend indicator values (TIV) system, which records sales at the point of delivery rather than the point of contract and measures the transfer of military resources as opposed to financial transfers – show no particular trend of increasing UK imports from the wider world or from the US specifically.[28] Indeed, since the end

Table 9: Conventional Arms and British Defence Imports, 2007–10.

Weapons Category	US Reported Exports to UK	UK Reported Imports from the US
Main battle tanks	–	–
Armoured vehicles	39	256
Heavy artillery pieces	–	1
Combat aircraft	3	4
Attack helicopters	19	15
Large surface ships	–	–
Guided missiles (excluding ground-to-air) and UAVs	973	823

Source: UN Register of Conventional Arms.

[28] 'The TIV is based on the known unit production costs of a core set of weapons and is intended to represent the transfer of military resources rather than the financial value of the transfer'. See SIPRI, 'Explanation of the TIV Tables', <http://www.sipri.org/databases/armstransfers/background/explanations2_default>, accessed 11 October 2013.

of the Cold War, imports apparently peaked in the period 1991–93. This also does not include UK spending on the JSF.[29] However, one clear message from the SIPRI figures is that almost 80 per cent of UK defence exports have come from the US. Just 17.5 per cent came from other European NATO and EU countries combined.

The German Approach to Acquisition

The urgent need for a fundamental change in the German acquisition system was highlighted by Minister of Defence Thomas de Maizière in a speech to the German parliament on 7 September 2011:[30]

> It is not a secret that the procurement process must be improved substantially. The processes take too long, and delays and postponements bind equipment, which is potentially not required anymore at all or only in reduced numbers.

The revised procurement and in-service process – the Customer Product Management (CPM) process outlined in Chapter II – offers at least three important innovations which will make the procurement process more accessible to industry, facilitate more effective business conduct, and increase industry's planning and investment security. First, through the integrated project teams (IPTs) which conduct each phase of the revised procurement and in-service process, the defence industry becomes an integrated component of the acquisition system – as far as is legally permitted and functional. Although the concrete mode of industry's participation in the IPTs has not yet been determined, their consideration as natural IPT members, particularly in the first part of the analysis phase, prior to the drafting of the Capability Gap and Functional Requirement (FFF) document, is a significant improvement. Through risk-reduction studies, industry could provide expertise to the IPTs regarding a project's technological, financial and operational risks, thereby gradually reducing its own commercial and reputational risk in a future realisation phase.

Second, the codification of a 'design freeze' by the Target Agreement (ZV) will substantially support industry in limiting technical, commercial and reputational risk, as it will prohibit the dilution of an agreement through subsequent changes and amendments. Lifting the design freeze is only possible if operational experience exceptionally demands a revision of the requirements or if disturbances to the project's

---

29 The authors are grateful to Siemon Wezeman of SIPRI for his clear explanation of what SIPRI seeks to capture and how it is measured.

30 See 'Rede des Bundesministers der Verteidigung Thomas de Maizière vor dem Deutschen Bundestag zum Etat des Bundesverteidigungsministeriums am 7. September 2011' (Translation by the authors).

progression result in violations of the ZV's tolerance limits. However, modification of the ZV requires approval by the departmental CPM Steering Board (CPM-LA) and, if applicable, a revision of the selection decision by the chief of staff, Bundeswehr.[31]

Third, the chief of staff, Bundeswehr's freedom to make his target agreement from at least three proposed alternative materiel solutions with a graduated degree of fulfilment of the functional requirements also has a positive effect on industry's business conduct. In the first place, it enables it to engage with a better-informed customer, that has a more coherent understanding of the technological and financial risks associated with a 100 per cent solution compared to, for example, 90 or 70 per cent solutions. At the same time, a selection decision for a less than 100 per cent solution should also translate into reduced technological, commercial and technological risk for companies, improving the timely and on-budget delivery of products and services.

In Germany, notwithstanding the lack of actual experience with the revised acquisition systems, this seems to provide a well-reasoned acquisition framework that more sensibly distributes financial and reputational risk between the demand and supply side, better ensures transparent and value-for-money tendering processes, and offers a higher degree of accessibility for both companies with an established footprint in the defence sector and companies eager to enter this market.

It is important to note, in this context, that the revised CPM procedures will present German-made defence products with increased competition from 'commercial, off-the-shelf' (COTS) products. Whereas in the past, no detailed assessment of COTS solutions to close identified functional capability gaps was necessary, the revised CPM specifically requires the IPTs to make such detailed evaluations as part of the CPM's analysis phase. Although the IPTs are also required to include defence industrial consideration in their assessments, a selection decision favouring a domestic solution will only be made if it objectively offers, on balance, a comprehensive case compared to a COTS solution.

The US Approach to Defence Acquisition

The US acquisition cycle for defence is, without doubt, complicated and systemic, supporting, amongst other things, a strong and healthy advisory and lobbying sector centred predominantly on Washington, DC. Coupled with the US constitutional separation of powers between the executive, legislature and judiciary branches, and a distinct governmental approach to budgeting, engaging with and understanding the US procurement

[31] See Robert Trice, 'Globalisation in the Defense Industrial Base', briefing to the Defense Science Board, Washington, DC, 11 December 2006.

model is not without its challenges. Indeed, for some, the very nature of defence acquisition in the US is a barrier to entry into the market.[32]

As has been stated frequently throughout this text, the defence market within a state is predicated on a niche and highly specialised, interdependent relationship between the buyer and seller. In the US, the DoD is the sole buyer in a monopsony of goods and services provided by only a few specific suppliers critical to each domain within the sector, such as land or maritime systems. Moreover, the domestic market is highly regulated and transparent, through congressional scrutiny and an alert and engaged lobbying community. This context makes for a highly specialised – perhaps even unique – acquisition process where contracts placed by the US government overtly sustain a specific defence industrial capability. Indeed, before exports from the US are conceptualised or included in the analysis, it is the only level within government capable of fulfilling this function.

Due to the fact that the normal dynamics of a free market are not a feature of the US defence ecosystem, laws, regulations, oversight controls and audits have evolved within a tight management system designed to enwrap defence acquisition, thereby ensuring – to American eyes – the kind of efficiency and 'competitive' fairness envisaged by classic market economists. In 2007, the US posted more financial auditors to Afghanistan than contracting personnel whilst, at the same time, 30,000 auditors were employed in the DoD to oversee the relationship between government and industry, played out through the defence acquisition process.[33]

Significant and voluminous federal congressional legislation contained within Federal Acquisition Regulations (FARs) and Defense Acquisition Regulations (DARs) detail exactly how business is to be contracted between government and the defence industrial base in the US. These regulations address the specific nature of accounting systems within companies doing business with the DoD, as well as quality-control procedures, health-and-safety protocols, project-management standards, governance, drawings and soldering techniques, all the way down to standards of cleanliness within rest areas on defence industrial sites. Defence businesses are regularly reviewed and evaluated against these standards to the extent that these form part of the drumbeat of the defence acquisition process in the US.[34]

---

[32] Author's interview with a senior US defence industrialist, 6 November 2012.

[33] Lieutenant General Ross Thompson, testimony before the Senate Armed Services Committee, 24 January 2008. See also Commission on Army Acquisition and Program Management in Expeditionary Operations, 'Urgent Reform Required: Army Expeditionary Contracting', 31 October 2007.

[34] See Gansler, *Democracy's Arsenal*, p. 157.

The government is intimately involved in the operation of the defence market. It controls almost all R&D, provides most of the money-for-progress payments, and provides much of the critical plants and equipment. The government's involvement in the day-to-day operating detail of the firms is so great that the defence market becomes totally unique and ceases to be a market in any traditional sense.

The defence acquisition process itself comprises a number of processes. It is, at one level, a requirements process, a contractual and procurement process, a production process, a congressional process and a budgetary process. All of these activities collude to produce a deliberate and systematic defence acquisition process, absolutely bespoke to the US.

The budgetary process is top-down through government, flowing from the president's priorities for defence. The Office of Management and Budget oversees executive agencies' allocations, which are approved and assured through congressional oversight. The DoD works through a series of five-year plans, with budgets for activities and programmes endorsed against specific departmental higher objectives. Specific capability requirements are lodged within this budgetary process, so that government resources, overseen by Congress, are aligned to specific defence objectives within overt spending and performance parameters and cost limits. This plays out in the public domain so that the spending profiles associated with each programme or project are visible to multiple stakeholders, including, of course, industry. Whilst the acquisition process in the US is complicated and multilayered, it is also remarkably visible and open to review and analysis. Yet the added cost of this regulation and oversight is estimated to be in the region of 10 to 50 per cent when compared to conventional commercial practices.[35] Moreover, the scale of the acquisition operation is simply staggering. Within a typical year, close to 4 million procurement actions may consume $284 billion of the defence spend, with 80 per cent of this going to American prime contractors.[36] Accountability, accessibility and procedural compliance, therefore, appear to be key features of the US government's sponsorship of the defence industry. These conditions reside, however, in a highly bureaucratic and bespoke context.

Imports and Exports

A common argument made in the US for government's involvement in defence industry is the politician's and official's support for business exports on behalf of indigenous companies. This rational discourse is

[35] See Mark A Lorell and John C Graser, *An Overview of Acquisition Reform Cost Savings Estimates* (Santa Monica, CA: RAND, 2001).
[36] Figures quoted for 2006. See DoD, 'Summary of Procurement Awards 2005–2006', 2006.

usually undertaken through narratives of domestic jobs and imperatives of securing the future of the workforce, balance-of-payment interests and the importance of projecting notions of soft power.[37] This is part of the script, but not the complete story in the three countries under consideration.

The United States

The US is the largest defence market in the world and companies such as BAE Systems, headquartered in London, believe that they will be more successful in doing business with the US government if they are based in America – hence the purchase of US-based businesses by BAE Systems. Likewise, US companies such as Boeing base themselves in other countries to access those markets, thereby growing their order books and balancing market risks and volatilities. It is a two-way street, and a common corporate strategy of expanding into new markets. In relation to US exports, the government is keen to enable overseas sales, providing key technologies are not transferred or compromised and the domestic customer preference is not lost. It is not a policy dogma in the US (unlike in the UK, perhaps), but is more a case of policy pragmatism.

Germany

As exports have become a central element of many defence companies' commercial viability, mitigating decreasing demand in traditional home markets,[38] the German defence industry requires substantial sponsorship by the German government to successfully exploit the highly competitive global defence market.

As part of its defence export policy, the German government applies four central sponsorship tools. For both civil and military exports, the government provides export credit insurances – so-called 'Hermes Bonds' – to secure export deals against customer-default risks, which could not be cost-efficiently covered by the private sector alone. In the defence domain, this transfer of financial risk from the private to the public sector improves the commercial viability of defence exports and, in particular, facilitates exports to 'third countries' in emerging markets with a comparatively lower financial standing.

---

37 See Per Lundin, Niklas Stenlas and Johan Gribbe (eds), *Science for Welfare and Warfare: Technology and State Initiative in Cold War Sweden* (Sagamore Beach, MA: Watson Publishing, 2010).

38 See Bundesministerium der Verteidigung, 'Minister de Maizière billigt Umrü stung', 21 October 2011, <http://www.bmvg.de/portal/a/bmvg/!ut/p/c4/NYuxEsIgEET_iIPGOHZiGgsb44zGjhCGnBJgLpfY-PFC4e7MK_bNwhNKo9nQG8YUTYAH9BYPw0cM8-bFK61UVrGgnRxNDnnJKSDjG-71OjphU3RcyS4yFnoynEjkRByqWYmKEThCL1WrpZL_qO_-cjvpbtc07VlfIc_z8Qec5KgJ/>, accessed 15 October 2013.

In 2011, Hermes Bonds valued at €2.5 billion in total secured defence export deals with Turkey (€2.5 billion), Pakistan (€11 million) and Peru (€6 million).[39] From January to November 2012, the German government issued six of these bonds with a total value of around €3.3 billion, including approximately €1.05 billion for the export of *Dolphin*-class Type 214A conventional submarines to Israel (€700 million) and Egypt (€405 million), built by Howaldtswerke-Deutsche Werft (HDW).[40]

Another form of direct sponsorship for German defence exports is the provision of military personnel for the demonstration of defence equipment's capabilities in the importing country and the training of the recipient country's military to operate the exported equipment. For example, in July 2012 the Bundeswehr sent one of its officers to Saudi Arabia to ensure firing safety during a demonstration of the Leopard 2 main battle tank, with Riyadh seeking to acquire the latest version, the A7+.[41]

In addition to this direct support for specific export deals, the German government plays an important role in companies' business-development efforts in the global defence market through the Bundeswehr's participation in defence exhibitions in Germany and around the world, as well as through the opportunity for company representatives to join official government delegations on their global travels.

As an illustration, consider the Bundeswehr's activities during last year's ILA Berlin Air Show. At a cost of €820,000 in total, the BMVg assigned 316 military and civilian staff to help industry promote its products and services. These activities included displaying equipment like the Eurofighter, NH90 and CH-53 helicopters, and the Luna remotely piloted system, as well as arranging 'courtesy visits' with foreign military representatives as potential customers of German defence products and services.[42]

The accompaniment of official government delegations to other countries is another common practice of sponsorship for German defence exports, as documented by the response of the German government to a parliamentary enquiry in May 2012. In the period October 2009 to April 2012, defence-industry representatives from companies such as Thyssen-Krupp Marine Systems, MTU Aero Engines, EADS/Cassidian, Atlas

[39] Gemeinsame Konferenz Kirche und Entwicklung, 'Rüstungsexportbericht 2012 der GKKE', Berlin, 10 December 2012, p. 7.

[40] *Spiegel Online*, 'Saudi-Arabien will erneut deutsche Panzer kaufen', 2 December 2012.

[41] *Spiegel Online*, 'SPD kritisiert Bundeswehr-Hilfe für Panzer-Test', 6 July 2012.

[42] See Deutsche Bundesregierung, 'Förderung von Rüstungsgeschäften und militärische Nachwuchsgewinnung auf der Internationalen Luft- und Raumfahrtausstellung 2012', Bundestag Drucksache 17/10705, 14 September 2012, p. 2.

Elektronik and Ferrostaal took part in twenty-nine foreign trips by the German chancellor, foreign minister and minister for the economy and technology to countries including Saudi Arabia, the UAE, Qatar, Bahrain, India, Singapore, Kenya, Nigeria, Angola, Argentina and Brazil, as part of the larger German industry delegation.[43]

While on the one hand these forms of defence export sponsorship are highly controversial within German society and politics, on the other hand, the German government's current sponsorship role is quite limited in comparison with other European countries like the UK, France and Spain, as well as non-European countries like the US and Russia. Unlike Germany, these countries all provide firms in their defence industrial base with considerable political and economic support in their export activities, embedding their defence export policies within the broader framework of their national foreign, security and economic interests.

The absence of a level playing field – both at the European and the global level – severely restricts the competitiveness of the German defence industrial base and presents it with a potentially existential strategic challenge. Against this background, the Federation of German Security and Defence Industries (BDSV) has urged the German government to advance its support of companies' export activities through:[44]

- The strategic and cross-government positioning of the German security and defence industry within the framework of German economic, foreign and security policy
- The creation of a federal cross-governmental defence export support organisation, in order to accelerate governmental co-ordination processes
- The simplification and acceleration of export-market access through support for international governmental agreements
- The establishment of an institutionalised training support capability in the German federal armed forces for export customers
- The provision of attractive, state-guaranteed terms of finance for export sales
- The acceleration of the licence-issuing process governed under the War Weapons Control Act (KWKG) and the Foreign Trade Act (AWG).

German Chancellor Angela Merkel explained in her speech at the 2012 Bundeswehr Convention that 'if you feel an obligation to undertake

[43] See Deutsche Bundesregierung, 'Auslandsreisen von Mitgliedern des Bundeskabinetts unter Begleitung von Rüstungslobbyisten', Bundestags Drucksache 17/9854, 31 May 2012.

[44] BDSV, 'Sicherheit made in Germany: Zeit für Veränderungen – Chancen erkennen und nutzen', July 2010, p. 15.

peacekeeping but cannot take on an active peacekeeping role everywhere in the world, you are also called upon to support reliable partners in their efforts to undertake such tasks', arguing further that 'other nations and regional organisations should not only be encouraged, but should also be enabled to solve specific conflicts themselves through training and equipment support'.[45] As such, Merkel echoes one of her previous speeches, in October 2011, in which she stated that 'it is generally not enough to send other countries and organisations words of encouragement if Germany shies away from military intervention', instead suggesting that Germany must also provide the necessary means to those nations that are prepared to get involved, emphasising that this includes arms exports.[46]

This approach to defence export policy faces rejection by large parts of the political opposition in the German parliament. The shape of the coalition government that is formed in the wake of the 2013 federal elections could, accordingly, be regarded as a key determinant of the future of German defence export policy.[47] Both Social Democrats and Greens have made clear their intention not only to align their defence export policies much more closely with the human-rights criteria of the political principles, but also to enhance both transparency and the role of the Bundestag in the German defence export-control system. Most likely, this will also affect political sponsorship for German defence exports, particularly to third countries.

The United Kingdom

Likewise, a significant element of government sponsorship of the defence industry in the UK relates to the government's active backing of defence exports. A key element in this policy stance came as early as 1966 when the Labour government established the Defence Export Services Organisation (DESO) to help identify market opportunities in the wider world and to support companies, in part with information, to help them to make sales. DESO was an organisation within the MoD, where there was a separate office which sought to co-ordinate DESO's exporting efforts with the UK's export-control system. This organisation was widely viewed as successful, providing effective help, particularly for medium and small companies that could not easily afford to monitor markets across the

---

[45] Angela Merkel, quoted in *n-tv.de*, 'Instrument der Friedens-sicherung: Merkel wirbt für Rüstungsexporte', 22 October 2012, translation by the authors.

[46] See Henrik Heidenkamp and Ferdi Akaltin, 'Confronting the European Defence Crisis: The Common European Army in Germany's Political Debate', *RUSI Journal* (Vol. 157, No. 2, April/May 2012).

[47] At the time of writing, German political parties are in the midst of post-election coalition negotiations.

world, and even surviving the Thatcher government with its aversion to any form of government industrial policy.

Under former Prime Minister Gordon Brown, DESO was renamed the Defence and Security Organisation (DSO) and its base transferred from the MoD to the Department of Trade and Industry, which became the Department for Business, Innovation and Skills (BIS) in 2012. There was concern that it would lose capability and effectiveness as its staff numbers were cut. Moreover, the change seemingly reflected reluctance, in parts of the Labour Party, for government to be explicitly involved in supporting arms exports, with the political risks that were inevitably involved. Defence exports were not universally popular in British society; the Campaign Against Arms Trade was a small but highly vocal group.

However, on coming to power in 2010, the current coalition government expressed an enhanced commitment to supporting defence exports, not least through ministerial time and attention, where appropriate. As in other states, exports were seen as an important element in assuring the survival of national defence businesses and the government was ready to do what it could to assist. Although the DSO was based in BIS, it continued to be able to use seconded MoD officials and serving military personnel, and to arrange the contact with defence ministers that foreign governments frequently sought. As a sign of the times, the government appointed a junior minister for whom exports were a particular responsibility and the prime minister himself opened the Farnborough Air Show in 2012. From 2012, the DSO also acquired responsibility for the implementation of the government's Industrial Engagement Policy to attract foreign investment to the UK.[48] The public presentation of this policy stressed the attractiveness of UK industry and capabilities, and did not refer to any need to invest in the UK by non-EU companies hoping to win British defence business.

Despite the survival, and success, or DESO and, later, DSO, it has not always been stated government policy proactively to help industry to flourish. Under former Prime Minister Thatcher, the British government concluded that it should not seek to identify future industrial areas where British business could be particularly successful ('picking winners') and that it did not matter which sectors led the drive for prosperity: there was no endorsement of the particular importance of manufacturing. However, the splits in government caused by the preference for selling Westland

---

[48] For details of DSO's structure and roles, see UK Trade and Investment (UKTI), 'How We Can Help – UKTI DSO services', 2 August 2013, <http://www.ukti.gov.uk/defencesecurity/item/107130.html>, accessed 11 October 2013; for information about the Industrial Engagement Policy, see UKTI, 'Industrial Engagement', 2 April 2013, <http://www.ukti.gov.uk/defencesecurity/defence/industrialengagement.html>, accessed 11 October 2013.

helicopters to Sikorsky of the US showed that not all ministers were happy to allow the private sector to make all of the government's own decisions. The subsequent Labour administrations did not markedly alter this stance, although, as noted, there was some recognition of the link between the ability to use the armed forces and the supply base which underpinned them.

The financial tribulations that began after 2009 led to a rather different stance. While the government could leave manufacturing industrial decisions to the private sector (and see many companies decline), it could not leave the financial sector to suffer the consequences of the decisions of bankers and others. Billions of pounds were spent saving major companies in the financial sector. The dependence of the British economy on financial businesses began to look somewhat risky, and there was certainly government readiness to recognise the contribution and potential of the high-technology industry in general and the aerospace industry in particular.

Most areas of the ensuing government initiatives focused mainly on the civil sector, for instance, the Aerospace Growth Partnership (which did have some concern with dual-use technology), but the defence sector should certainly benefit from the commitment of government and industry to fund 500 Master's-level degree places in aerospace engineering from 2012 to 2015. The principle of joint funding between government and industry was also apparent in the commitment to invest £60 million of government money in a UK Centre for Aerodynamics and £40 million in a Rolls-Royce-led programme to develop low-carbon engine technology.

Although past governments have paid lip service to the idea that UK requirements should be written so that the equipment subsequently developed will be attractive to export customers, this point was given new emphasis by the coalition government at an early stage of its tenure. There were, however, particular challenges associated with implementation, notably the questions of who could speak authoritatively about what would do well on world markets and at what stage in the requirements-definition process these considerations would be fed in. There was also the question of how this consideration, which implied a UK readiness to develop its own equipment either alone or with partners, could be reconciled with the headline statement from the *National Security Through Technology* White Paper that the default position of the UK would be to buy defence equipment off-the-shelf on the global market. The major project that quickly emerged as the test case for this approach was the Type 26 warship, for which the UK sought to attract partners to commit to the project from an early stage. That effort had not borne fruit by mid-2012, but the hope remained that the modular approach to design would enable a flexible product that could be customised for different countries. Looking forward, much will depend on the readiness of the single services to contemplate allowing export considerations to influence their requirements.

The emphasis on exportability did, however, send a clear message to the UK services that they should also play their part, in particular, by offering equipment demonstrations and making sure their training programmes would be available, if necessary, to foreign operators of British equipment. Under the Labour government, these issues had been a matter of some dispute.

Defence Industrial Strategies

Not all states have a defence industrial strategy. In the US, there is a long tradition of federal government possessing a stated strategy for the health of its defence industrial base, reviewed every five years. The Germans, by contrast, as will be shown below, would not conceive themselves to possess something that could be defined as a strategy for defence industries. The UK, in yet another approach, has moved from a shopping list of capabilities articulated in its 2005 Defence Industrial Strategy to a more nuanced approach. What is not in doubt is that, across much of the Western world, there has been a propensity for governments to enter into so-called long-term partnering contracts with private-sector organisations for the delivery of defence capabilities.[49] In the US, the trend has been more concerned with teaming between defence businesses before overt engagement with government, either in response to a request to tender or in discussions on future requirements.

As part of the inherent sponsorship of defence industries across the US, the government has encouraged suppliers to team together. For a specific requirement this can have the effect of creating a monopoly which offers immediate commercial benefits and assurances to the businesses in question. It keeps them in the market for future engagement, but also seems to imply that government loses the benefit of competition on the current programme. That government is willing to forego this advantage hints at the importance to the political elite of keeping the industrial base in the US at a certain mass. For example, in 1997 Newport News Shipbuilding and General Dynamics Electric Boat Company formed a team to design and manufacture advanced submarine systems.[50] The cost

---

[49] See John Louth, 'The Sum of its Parts? Partnering, the MoD and Industry', *RUSI Journal* (Vol. 157, No. 2, April/May 2012).

[50] See 'General Dynamics and Newport News Shipbuilding Sign Teaming Agreement for Construction of New Attack Submarine', press release, 25 February 1997, <http://www.generaldynamics.com/news/press-releases/detail.cfm?customel_dataPageID_1811=13058>, accessed 15 October 2013.

per unit was higher as a result, but the DoD was said to be content as both businesses stayed afloat in the maritime market.[51]

Teaming partners have noted that there is typically insufficient demand to support two leading companies per defence sector in the US, so that if government wishes to keep both principal businesses in play, this potential cost growth is a legitimate use of public money and a cost of sponsorship. This is an argument, of course, that would be hard to promote in Europe, but hints at how seriously the notion of sponsorship is taken in the US. These types of acquisition solutions have been followed in the UK, for example, by teaming or partnering arrangements for test and evaluation services, as with the Long-Term Partnering Agreement between the MoD and QinetiQ, and the partnership between the government and BAE Systems for the production of small-arms ammunition. Industry and government, therefore, work hand-in-glove to deliver services through partnering arrangements as a cornerstone of deliberate – or emergent – defence industrial strategy.

German Defence Industrial Strategy in the Making

Fundamentally, the German government has not yet developed a coherent defence industrial strategy which functions as a reference point for itself and for industry, generates a coherent cross-government understanding of defence industrial policy, and gives the private sector – particularly original equipment manufacturers (OEMs) and SMEs – a proper sense of planning or investment security.

As outlined in Chapter II, the German defence industrial strategy is codified in various documents, including the recently amended CPM, valid from January 2013, the 2007 BMVg–BDI Joint Declaration on Core National Defence Technological Capabilities, the 2006 White Paper on German Security Policy and the Future of the Bundeswehr and the 1999 Framework Agreement on Innovation, Investment and Cost Effectiveness in the Bundeswehr. Such documents do not, however, present a clear sense of the German government's defence industrial strategy. Neither are they structured in a transparent document hierarchy, nor are they sufficiently aligned with one another. Furthermore, they offer mostly generalisations, which do not capture the significant changes in the defence industrial landscape witnessed over the last decade and leave significant room for interpretation with regard to the orientation of the future equipment programme and the determination of core national defence industrial capabilities. In combination with decreasing demand in the German domestic market, this lack of strategic clarity generates substantial uncertainty for the German defence industrial base.

[51] See Gansler, *Democracy's Arsenal*, p. 168.

In order to become a truly beneficial component of the German government's approach to defence industrial sponsorship, it is crucial that any future German defence industrial strategy document goes considerably beyond these generalisations and formulates pronounced policies regarding, *inter alia*, the future relationship between government and industry, the contribution of the defence industry to military operations, the changing boundaries and limitations of the private sector in defence, future requirements and programmes, and the skills and competencies required in the public and private sectors.

US Defence (Industrial) Strategies

As a means of addressing the risks and vulnerabilities that exist globally, the US perceives a capable and secure on-shore defence industrial sector to be a strategic national asset, which contributes significantly to the slippery concepts of US political power and military might.[52] As Ashton B Carter, the former US under secretary of defense for acquisition, technology and logistics, stated: 'A strong, technologically vibrant and financially successful defence industry is … in the national interest. In this respect the warfighter's and taxpayer's interest are fundamentally aligned with … industry.'[53]

Driven in part by this policy insight, as well as by a growing US fiscal deficit, Carter and fellow US senior leaders have increasingly championed the promotion of a new American defence industrial strategy that will secure long-term innovation and productivity, offering solutions that are both profitable to the industrialist and economically efficient for the taxpayer.[54]

At the heart of this refreshed defence industrial strategy is a reliance on classic market forces to initiate enduring structural adjustments to the defence industrial base, whilst concurrently ensuring that the US government continues to sponsor defence businesses operating within its territory. The concept at the heart of this policy is that the confluence of global capital markets, technological innovations, intellectual creativity and government consumer confidence will deliver a sustainable and globally dominant defence industrial base, or so policy-makers believe. Moreover, as competition is thought to be the primary enabler of economic prosperity, US policy does not extend to the further

52 Heidenkamp and Akaltin, 'Confronting the European Defence Crisis', p. 10.

53 Ashton B Carter, 'The Defense Industry Enters a New Era', speech at the Cowen Investment Conference, New York, NY, 9 February 2011.

54 DoD, 'Better Buying Power: Guidance for Obtaining Greater Efficiency and Productivity in Defence Spending', memorandum for acquisition professionals, 14 September 2010.

consolidation or rationalisation of principal weapons systems, the nomination of prime contractors or other dominant industrial actors.

Significantly, as will be discussed, US defence industrial policy also notably avoids focusing on prime contractors, as US policy-makers believe that smaller firms and new entrants are liable to generate new solutions and refreshed technologies, thereby offering battlefield advantages and commercial efficiencies. This is especially important for service-sector companies, as about half of all defence contracts by value in the US are awarded to service providers, rather than to the more traditional defence equipment manufacturers. This focus on service companies as incubators of innovation is at the centre of the US commitment to review and revamp its industrial base, in each defence capability area in turn, from aircraft manufacturing to small arms and ammunition. When each capability area and sector is of the right shape and size, the US government is committed to developing policies, processes and commercial incentives to secure, sustain and protect them.[55]

It can therefore be asserted at the outset that, as a dominant global actor aware of the risks posed in an uncertain modern era, the US commitment to its defence industrial base, whether truly indigenous or adopted and absorbed, is part of an overt policy stance. It is believed in Washington that a strong national-security and defence industrial sector will ensure that the US remains vigilant and potentially ready to address unidentified strategic shocks, and therefore will be prepared for the unprecedented pressures that such shocks place on policy-makers and wider society.

The implementation of this policy initiative to support the full spectrum of military responses through reshaping the defence industrial base has begun under the Obama administration, with industry and commerce seen as critical national resources and de facto sources of power. As an aside, this also offers opportunities to non-US corporations, as Carter and his colleagues are committed to reducing barriers to market entry and, where appropriate, to embracing new entrants, both from within and outside the US.

The UK Response

Given the strong policy preference for a robust defence industrial strategy in the US, it seems sensible to assess how its close ally, the UK, has approached this topic. As noted previously, in the UK government's *National Security Through Technology* White Paper, released on 1 February 2012, the government led heavily on the requirement to buy

[55] Robert Gates, speech to NATO Strategic Concepts Seminar, National Defense University, Washington, DC, 23 February 2010.

defence equipment 'off-the-shelf' and through 'open competition'. Naturally enough, industry is used to selling its equipment and competencies to government through open competition, so this was a well-understood and expected component of the White Paper. The government's preference for purchasing equipment off-the-shelf, where possible, had also been widely trailed in its commentaries prior to the paper's release. The suggestion in the document that government may intervene in future to protect notions of sovereign capability and technological advantage drawn from the UK's defence industrial base seemed in line with the US stance and potentially offers a counterweight to the preferred off-the-shelf policy.

The paper, unsurprisingly, championed defence and security exports, with this supposedly being one of the major 'takeaways' from the government's consultation with British industry prior to the White Paper's release. The government's commitment to maintaining the science-and-technology element of the defence budget at 1.2 per cent per annum also suggested recognition of its importance to innovation and future defence capabilities. How this fixed element played into the notion of a balanced defence portfolio was not considered by the politicians, however, and there is some debate about the detailed processes for maintaining this level of investment year on year.

Significantly, the paper goes to great lengths to stress the importance of small businesses in the UK's defence and security effort. Indeed, with more defence-industry SMEs in the UK than in France, Germany, Spain and Italy combined, their economic significance was seen as critical to the government's plans to reduce the country's budget deficit. This is interesting policy, but it is not a coherent defence strategy. Stakeholders within the UK defence and security markets and policy domains were crying out for a refreshed defence industrial strategy. Indeed, government ministers appeared to promise that the White Paper would constitute such a strategy document. For example, at the Conservative Party conference in early October 2011, politicians and officials stated that the document would replace the 2005 Defence Industrial Strategy and provide clear guidance and sector prioritisation to industry. This, of course, has not occurred, and UK policy seems riddled with inconsistencies. For example, the government's commitment to championing defence exports, whilst laudable, is logically incoherent beyond the immediate term. Traditionally, advanced states develop their own defence equipment, prove it through deployment with their own armed forces, and are then able to sell it to others. If the UK government is to purchase its defence equipment off-the-shelf from the world market, the UK will not necessarily research and develop its own capabilities or equipment packages in future. Under this scenario it is difficult to fathom just what defence and security equipment the country will be able to export in the mid-to-long term.

In the UK, the *National Security Through Technology* White Paper was probably the best industry and other stakeholders could have hoped for at a time when the MoD's focus was on deficit reduction, reorganisation and redundancies. However, it represents a set of policies in marked contrast to what has been witnessed in the US, and even in European polities such as Germany. This matters greatly to the UK, if traditional notions of national defence and security capabilities are compromised, and the UK's ability to play the role it wishes in world affairs is lost. Indeed, if there is no enduring national defence industrial base, and only off-shore foreign suppliers, the ability of governments to undertake independent action is lost. This, at least, seems the lesson from the US policy stance, and is a cornerstone of the reasoning behind government's sponsorship of the defence sector within its borders.

Implications of Government Sponsorship

At one level of analysis, some may argue that government's sponsorship of an ostensibly private-sector defence industrial base means that these businesses have a privileged relationship with the government over other sections of the economy, enjoying, for example, beneficial funding for R&D. This may be so, but close sponsorship by government also means that defence businesses may not necessarily enjoy the freedom of action, strategic options or sources of finance open to other businesses in Europe or the US. A government's interest in its defence industrial base may therefore be a double-edged sword.[56]

The level of regulation associated with acquisition requests (FARs and DARs) has contributed to this sense of monopsony within the US defence industrial base and its relationship with government, as only those companies historically embedded within the defence ecosystem have the structures, procedures, personnel and mindset to work within the space. This is an enormous barrier to entry into the market, effectively keeping out many innovating smaller, niche businesses unless they go to market in the US through one of the larger prime contractors, which already systemically conform to government regulation of the sector. Consequently, innovation has to be nurtured by government through its dominance of the R&D space, since innovating individuals and businesses – which should form the base of the defence value-chain – are effectively regulated out of the market. Government is finding that it can sponsor programmatically, make funding available, and invest in defence R&D, but that it is hard to sponsor step-changing innovation or what can be described as a scientific

[56] Author interview with a senior US defence industrialist, 6 November 2012.

'eureka' moment.[57] For that, a more flexible, organic set of market and governance arrangements may be necessary as technological and information transformations continue to dominate the high modernity of the twenty-first century.

Sponsoring (for) Whom?

An increased focus on SMEs seems to have become important and timely. In both the UK and Germany, a narrative of implied preference has been witnessed for smaller, niche businesses to innovate within the defence market, with a number of policies implemented to try to bring this about. There is a similar, overtly stated ambition across the US polity for the same outcome. As has been seen, it is a fairly common transatlantic narrative that greater innovation from research investment is to be found in smaller firms rather than the large defence prime contractors. The evidence behind this view is somewhat ethereal, yet the certainty with which it is held by many prompts political and organisational responses. In the US, Congress established in the 1980s the Small Business Innovation Research (SBIR) programme, mandating that 2.5 per cent of all government-funded R&D must be contracted with small businesses. Each government agency, including those within the defence sector, was to annually articulate research areas that would benefit from innovation, seeking matching proposals from the specialist SME community. Moreover, private equity houses and venture capitalists often note, from public sources, which businesses win these contracts, ensuring that successful businesses have plenty of choice when it comes to accessing funds.

There is evidence to suggest that the championing of smaller businesses through mandatory contracting has led to successful and innovative technology solutions being made available on the US front line, justifying the US policy response of small business set-asides. Yet the worry is that bureaucratic initiatives, however well intended, can lock out the innovation that fails to perform to type, perhaps by being sourced and developed from within a prime contractor. Either way, the SBIR initiative seems to enjoy cross-party support and is a key feature of each individual country's stated or emergent industrial policy.[58]

This focus on smaller businesses sits within a broader narrative of who sponsors defence industries from across the US polity. It is not a

---

[57] See Nassim Nicholas Taleb, *The Black Swan: The Impact of the Highly Improbable* (London: Penguin, 2007).

[58] See the National Research Council of the National Academies, *An Assessment of the Small Business Innovative Research Program at the National Science Foundation* (Washington, DC: National Academies Press, 2007) and House Small Business Committee, 'Manzullo: US Small Businesses Secure Record Amount of Federal Prime Contracting Dollars', 21 June 2006.

simple question to answer. As has been discussed, the US federal government, with its separation of powers, sits astride an enormous national defence budget, a significant portion of which is used to contract with or, in other ways, support the efforts of defence and security businesses. At one level, therefore, the federal government sponsors the defence industrial base in the US. This is, however, only one part of the story and has to be viewed alongside the influence, for example, that the single services in America can wield, the power of regionalism and statism and the sheer bulk of the lobbying industry in Washington, DC and state capitals. It is therefore a very complicated picture indeed, with multiple drivers and causes, which acts as the perfect metaphor for government's sponsorship of this pivotal industrial sector.

A Final Word on Sponsorship

This chapter indicates that governments continue to sponsor their defence industries, but that the practice is contingent upon various drivers: national preferences, the politics of the moment, constitutional arrangements, and practical considerations such as affordability and the quest for ever-greater governmental efficiencies and economies.[59] A state develops its own system of sponsorship in response to these subjective and shifting factors. It has been demonstrated that the governments of the US, UK and Germany, as examples, have all developed policies and processes to sponsor defence businesses operating from within their respective borders, in varying degrees supporting or promoting the activities of these companies. However, it has also been shown that this act of sponsorship itself is essentially a national imperative, and varies in detail from state to state.

In the US, for example, the stated policy preference is to conceptualise on-shore defence industries as an essential component of the country's military instrument. Having a healthy defence industrial base across the US, so the argument goes, enables the country to possess and retain a military and technological advantage over rivals: the core elements of its defence posture. It ensures that the country is able to surge its economic might during periods of tension and conflict, and that skills and knowledge are pooled to generate the desired military effects. Significantly, the private defence sector in the US is perceived to represent the US arsenal, and therefore to be worthy of active government sponsorship and nurturing. Consequently, it is natural for US politicians and commentators to discuss and debate an active US defence industrial strategy to the

[59] See Alvin Toffler and Heide Toffler, *War and Anti-War: Survival at the Dawn of the Twenty-First Century* (New York, NY: Little Brown, 1993).

extent that its absence would be perceived as a major failure or abdication of government responsibility.

In Germany, paradoxically, there is a commitment to the notion of government sponsorship of on-shore industries through, for example, the development of an active defence industrial strategy, but this is driven by economic reasoning as much as national-security imperatives. As part of a highly pragmatic approach, German politicians seem to recognise that investment in the private sector generates economic benefits, meaning that the emerging German defence industrial strategy is also a component in the state's fiscal strategy and policies for economic and social development. In addition, much of the debate in Germany on defence industrial and acquisition reform must be seen through the prism of a wider narrative concerned with the purpose of Germany's military forces and the future defence and security stances open to the country. If Germany is to lead politically and economically in an expanded Europe, as the advanced world recovers from the global economic crisis of 2008, the question arises as to whether it will have to carry more of the international defence burden. This is the key defence and security question in Berlin (as well as the capital cities of its international partners), and answering it effectively may require Germany to look ever-more closely at the defence acquisition and industrial policies emanating from North America.

The UK, by contrast, since 2010 has had a public discourse that seems to favour global suppliers and open international competition over governmental patronage of a national defence industrial base. Given that the UK has been at war continuously since the beginning of this century (usually in support of the US) and is, in part, dependent upon these British businesses, this may appear somewhat surprising. Moreover, despite this narrative of international supply to UK forces, British politicians and officials continue to see defence exports from the UK to other states as a critical element of future economic growth and development. It has been stressed that, typically, defence goods and services must be operated and proven by a 'home' nation if they are to possess realistic export potential.[60] The UK's sponsorship of its own defence industries therefore appears somewhat conflicted, hinting at the inconsistencies and contradictions inherent in the state's endeavour to make sense of its sponsorship role.[61] Of course, governments seek to redress these inherent tensions through the active regulation of this sector. It is to this area of state activity that the following chapter turns.

---

[60] See David Gould, 'Procurement Reform', in Codner and Clarke (eds), *A Question of Security*.

[61] See Keith Hayward, 'Defence Industrial Strategy under the Coalition', in Codner and Clarke (eds), *A Question of Security*.

IV. THE REGULATION AND CONTROL OF DEFENCE BUSINESSES

This chapter turns to the third element of the defence industrial triptych: regulation. Specifically, it considers certain governmental control mechanisms that apply to firms supplying defence ministries and their armed forces with specialist goods and services (as opposed to items such as bottled water or diesel fuel that are essentially civil in nature, but are consumed by defence forces). This chapter analyses areas in which defence companies receive special treatment with regard to regulation governing the wider business sector, especially in relation to monopolies and mergers, and notes areas where defence firms do not enjoy special exemptions, including, for example, the law against corrupt practices and health-and-safety regulations. It also addresses the restrictions on hiring of former government and military personnel by the private sector, and touches upon anti-bribery and anti-corruption policies and practices.

Governments everywhere seek to regulate the businesses on their territory on issues such as minimum pay, terms of employment, pollution, health and safety, and anti-competitive behaviour, to name a select few. Multinational businesses must cope with, and even seek to exploit, the differences in approach between the governments relevant to their activities. However, the material in this chapter demonstrates that defence businesses have specific laws, regulations, policies and practices to address. These factors have a real impact on their ability to optimise the use of their intellectual and other resources.

Systematic government efforts to secure peacetime control over the activities of defence-equipment manufacturers on their territory have existed since the end of the First World War and the upsurge in public perception that firms profiting from arms sales were promoting conflict. Nevertheless, it took the real prospect of war with Hitler's Germany for the UK to bring in significant export-control legislation, in the form of the Import, Export and Customs Powers (Defence) Act 1939. After the Second World War, there was a unanimous view in the Western world that the advantages of trade, and even any notional right to trade, needed to be viewed in the light of possible negative security consequences. From the perspective of NATO, at least until the onset of the 'détente' period, a key element of the Cold War was the effort to restrict any growth in Warsaw Pact military power by the denial of useful Western products. The Coordinating Committee for Multilateral Export Controls was set up, under the leadership of the US, to make sure that Western goods and information which might directly or indirectly benefit military strength in the Warsaw Pact countries, and of course Communist China, remained unavailable to these states. The concepts of equipment with a specific military purpose and dual-use goods – which could be used for either civil or military purposes – became clearly established in governments' mindsets, although the effect of this definition in terms of actual products developed remains contested: sustained technological change and a thawing in East-West tensions did not reduce the difficulties involved.

The Control of Information

Formal legal systems for the protection of information significantly predate controls on exports and these early regulations have in many ways acted as a key enabler of both government controls over industry and of closer government–defence industry relations. The control of information critical to defence remains a hot topic as information is seen by governments and militaries as a battle-winning component, offering a distinct and identifiable advantage over an adversary. Controlling and protecting this defence information is perceived as a central government role.

The United Kingdom

The first version of the Official Secrets Act was passed in Britain in 1889 and has since been periodically amended and updated. This law makes it an offence for anyone subject to the Act to pass classified information to any unauthorised person. The Act is quite draconian in scope, committing those affected to its terms for the rest of their lives and not just for a single period of employment. Classified information can be passed only to another authorised person, who would also normally be a signatory to the Act, thus ensuring that such information remains under constant control.[1] However, there has been extensive debate in the UK as to whether a public-interest defence for revealing information to the public should be available, and whether only unauthorised disclosures which caused serious damage to national security could result in criminal prosecution.

[1] For a clear analysis and historical background, see Lucinda Maer and Oonagh Gay, 'Official Secrecy', Standard Note SN/PC/02023, House of Commons Library, 30 December 2008.

Some argue that disciplinary action should also address disclosures of lesser consequence.[2]

The UK scheme for information provides for classification at four main levels: Restricted, Confidential, Secret and Top Secret, with written guidance being available as to relevant criteria for each category. There are rules for how information at different levels should be protected, both when it is being stored and when it is being transferred. In 2014, the UK will move to a three-tier system of classification (Official, Secret and Top Secret), which is meant to be more suitable for the electronic communication age.

This classified-information structure sits alongside another system for the clearance of individuals, which specifies what sort of information they can receive. There are three normal clearance levels for personnel: Basic, Security Check (SC) and Developed Vetting (DV). The Security Check category means that, when appropriate, an individual can be given information which is classified up to Secret. Developed Vetting authorises the individual for Top Secret clearance. Each level of clearance involves a different level of investigation into an individual's background, with DV clearance requiring a somewhat intrusive investigation into an individual's private life. The government's obvious concerns appear to be about possible sympathies with a potential adversary, lack of respect for the law, vulnerability to blackmail, and financial difficulties.

The responsibility for investigating individuals and assessing their suitability for clearance is entrusted to the Defence Business Services National Security Vetting (DBS NSV – formerly the Defence Vetting Agency), which deals with civil servants, military personnel and private-sector staff. This is a governmental body, in contrast, for example, to the US, where some vetting activity is contracted to the private sector.

Companies that seek or require access to classified information must have relevant staff cleared by the DBS NSV, and they and their premises must obtain List X status. This essentially requires the organisation, including its infrastructure, to go through an investigation to show that it can and will protect, according to government rules, all of the information that is entrusted to it on a specified site or sites. Formally, List X status is given to a particular site, rather than to the company as a whole. Companies that have obtained List X status for one or more of their sites normally have a security officer to liaise with government authorities regarding personnel (including potential new employees and any problems with existing staff), equipment and infrastructure. The government clearly asserts that non-List X companies can be issued with invitations to tender and can even be selected as contractors. However, they must

2 *Ibid.*

become List-X qualified before they can be awarded a contract involving classified material.[3] A further element of the UK control system involves government guardianship of information that is of particular value or sensitivity to industry. The classification category 'Commercial in Confidence' is used to cover bids and other documents involving industry, which means that information should not be passed to other commercial bodies. Information about company prices, costs, contract terms and, sometimes, technology are covered by such arrangements. In defence acquisition, governments need to assure firms that they will not abuse private intellectual property and the commercial-in-confidence arrangements do much to promote this. A government that wants companies to provide access to their ideas must give assurances that those ideas will be protected. However, this is a problem that has not yet necessarily been solved by the UK Ministry of Defence (MoD): government officials holding information from the private sector may be tempted to use it elsewhere if they feel it will enable them to get a better price or performance for the government.

An additional marking applied to some documents is 'UK-eyes only', which means that information is not to be exposed to any foreign national.

The United States

In line with the practices of most advanced industrial societies and their military components, the United States employs a series of security classifications, caveats, constraints and processes to protect its military and industrial expertise and technologies. This takes the form of the classification of information, security clearances applied to private-sector staff and robust government direction of standards relating to defence sites, whatever their ownership. This means that government can overtly direct the behaviours and practices of the industrial and services base by requiring compliance with governmental specifications and norms. It is not possible to act as a defence contractor without embracing these security standards as they are backed by federal and state statutes. The protection of defence information is specifically guaranteed under the US Espionage Act of 1917 and other federal legislation, but in the US there is extensive if intermittent debate about the legal arrangements to protect classified information in general, and even governmental information as a whole. There is particular concern surrounding how controls of government information not relating to national security interact with the freedom of speech and expression assured under the First Amendment

[3] See Cabinet Office, 'List X Contractual Process', version 3.0, October 2009, paras. 14–18.

(in Britain, the challenge is to reconcile the Official Secrets Act with the Freedom of Information Act). The US system for classifying information is similar to, but subtly different from, that of the UK – there is no Restricted level, but non-classified information may be marked 'For Official Use Only' or 'Sensitive but Unclassified'. The US can also designate some material as 'No Foreign Nationals' (NoForN).

Whilst US regulations in this area are primarily the preserve of the federal government, the nature of the distribution of powers in the US means that specific state regulations can also bind federal employees and industrial staff to specific protective regimes. This means that the burden of compliance can vary from state to state, ensuring that legal practitioners and commercial officers enjoy significant hegemony over the system.[4]

Germany

Germany has an extensive, though bespoke, legal and regulatory system for the designation and protection of classified information. On first glance, it may seem very similar to the UK system, but with the Federal Ministry of Economics and Technology (BMWi) trade and economy ministry in a more prominent position. For example, Germany has an equivalent of the UK's X-listing mechanism and similar arrangements to facilitate the entry of new firms into the defence sector by allowing them to bid for work once all the appropriate security arrangements are in place. However, on closer examination, the German system is more bureaucratic and burdensome when compared to those of its allies.

In Germany, classified information (*Verschlusssachen* or VS) comprises facts, matters or knowledge that, regardless of form, need to remain confidential to protect the public interest. Companies active in the defence sector may frequently require access to such classified information initially to bid for public tenders and eventually to deliver products and services procured by the BMVg. The protection of all classified information is governed by legislation.[5] Furthermore, the protection of classified information by BMVg contractors and sub-contractors is governed by German law reflecting European requirements.[6]

[4] See David M Walker, *DoD Transformation Challenges and Opportunities* (Washington, DC: Government Accountability Office, November 2007).

[5] See 'Allgemeine Verwaltungsvorschrift des Bundesministeriums des Innern zum materiellen und organisatorischen Schutz von Verschlusssachen (VS-Anweisung – VSA) vom 31. März 2006'.

[6] 'Vergabeverordnung für die Bereiche Verteidigung und Sicherheit [VSVgV] zur Umsetzung der Richtlinie 2009/81/EG des Europäischen Parlaments und des Rates vom 13. Juli 2009 über die Koordinierung der Verfahren zur Vergabe bestimmter Bau-, Liefer- und Dienstleistungsaufträge in den Bereichen Verteidigung und Sicherheit und zur Änderung der Richtlinien 2004/17/EG und 2004/18/EG', para. 7.

There are four classification levels issued by official classification authorities: 'Top Secret' (*Streng Geheim*) if knowledge of the information by unauthorised persons would endanger the existence or essential interests of the Federal Republic of Germany or of one of its states, 'Secret' (*Geheim*) if its knowledge by unauthorised persons would endanger the security of the Federal Republic of Germany or one of its states or do substantial damage to their interests, 'VS-Confidential' (*VS-Vertraulich*) if its knowledge by unauthorised persons would harm the interests of the Federal Republic of Germany or one of its states and 'VS-Only for Official Use' (*VS-Nur für den Dienstgebrauch*) if its knowledge by unauthorised persons could have a negative impact on the interests of the Federal Republic of Germany or one of its states. These are, of course, broadly comparable with UK criteria and thinking.

The BMWi will issue security clearance to bidding companies if their legal representatives hold security clearance, if they have a qualified 'security officer' (*Sicherheitsbeauftragter* or SiBe) and at least one deputy, if they have been nominated and instructed by the BMWi on their obligations and, at this point, if required materiel confidentiality measures have been enacted. In particular, the BMWi's security clearance contains restrictions regarding additional confidentiality measures to be implemented by the company before the actual execution of a confidential contract.

Furthermore, bidding companies and their potential sub-contractors must provide 'declarations of commitment' regarding the protection of all classified information in their possession, to which they are bound for the full duration of the contract as well as after its cancellation, dissolution or expiration. If the application for bidder status or the preparation of an offer requires access to confidential information classified 'VS-Confidential' and above, companies must uphold the BMWi's security decision and provide declarations of commitments before being granted access to such information.

The assured compliance with these classified-information regulations through organisational, personnel and material measures comes at significant cost to companies active in the defence sector. However, the consequences of failure to comply with them are significant, at the corporate and also potentially at the personal level, ranging from exclusion from the relevant and future tendering processes and the imposition of fines to criminal prosecutions of companies as well as individuals.

In particular, small and medium enterprises (SMEs) with limited overhead resources depend on close co-operation with original equipment manufacturers (OEMs), contracting authorities and the BMWi in order to navigate the complex regulatory environment, which may also add to the substantial market-entry barriers in the defence sector.

However, the nature of defence products and services arguably makes the classification of information a *sine qua non* of governmental and corporate conduct, not only to correspond with the public interest in the protection of confidential matters, but also to ensure companies' sustained compliance with procurement and cartel law.

Multinational Systems

Obviously, an important question is how national and international systems interact. As far as international organisations are concerned – and there are a number which deal with Western defence and security in particular – member governments must decide on what information they will release to the international body and its membership, while each international body must have arrangements for the protection of information handed to it. For instance, NATO and the EU each have their own levels and titles for classifying information. Beyond this, governments have to make particular arrangements for bilateral and multilateral information-sharing outside the framework of an international body. Australia, Canada, New Zealand, the UK and the US have developed particular intimacy through the frequent sharing of intelligence information within the 'Five Eyes' framework, although this has only limited significance for defence industries.

These information- and personnel-control systems mean that defence firms, in general, face greater human-resource challenges than those in the civil sector, since they are forced to recruit from a smaller pool of labour. In an era in which most Western governments believe that the current educational and social system is not producing the number of engineers needed by the economy, defence companies must hire from a national market of people willing and qualified to be security-cleared: a Developed Vetting clearance in the UK involves an individual being willing to offer up details of their private lives, including sexual habits, and to submit to external examination. Furthermore, in an age of large-scale immigration, being a British or US citizen who was born overseas to foreign parents does not prevent the granting of high-level clearance, but it does not make it easy. Concerns about the incidence of industrial and security espionage against NATO states, including that perpetrated by other Western states, mean that clearance standards will not be easily lightened. One consequence of this is that major employers of engineers, such as BAE Systems and Rolls-Royce, actively seek to stimulate enthusiasm for their activities among young people while they are still in school. They also seek largely to 'grow their own' staff, rather than rely on outside recruitment for middle-ranking and even senior positions.

For defence companies that run development, manufacture or support operations in more than one country, and which may be seeking to develop offerings that could serve even wider markets, human-resource management hurdles become even greater. A company such as MBDA, for example, which ostensibly is a multinational business with a French-British-German-Italian identity, cannot freely move its technical expertise from one country to another. A British engineer facing a technical challenge may know there is pertinent expertise within the company in another country, but he or she is not easily allowed access to it. Multinational defence firms need government permission to move staff into the positions in which they can add greatest value if that involves crossing a border into a country where the individual lacks the necessary clearance. European firms such as BAE Systems, QinetiQ and EADS, which comprise significant US businesses, find it hard to use non-US nationals in these businesses because of the significant limits on the information that they can receive. Normally, a Western government will not give security clearance to a foreign national, although exceptions can be and are made, especially when European governments want access to US expertise. It is understood that US nationals are among the Lockheed Martin and Jacobs teams that are partners in the government-owned contractor-operated (GOCO) arrangement that runs the UK's nuclear-weapons plants.

It is important to appreciate what is involved in the ownership of a defence company in the US by a British firm. As a rule of thumb, only cleared US citizens have access to classified information. Normally, the DoD's Defense Security Service insists that the company is run by a 'proxy board' comprising only suitable US citizens, leaving the owning company free to repatriate profits but with little knowledge of the technology, projects or even details of the strategy of its subsidiary. When the US has a high level of confidence in the security arrangements of the owning company, it may allow a 'special security arrangement' (SSA) in which the US subsidiary board may have some foreign members (British members in the case of Rolls-Royce and BAE Systems).[7] However, when the board discusses matters concerning US classified information, the non-US citizens are required to leave. Due to the extra oversight that an SSA allows, it is a prized status in the US, which those who have it are loath to put in jeopardy.

The control of information, therefore, is subject to very specific national regimes of regulation but, as this chapter demonstrates, some common features can be identified. First, the practice of protecting state secrets and other sensitive information is highly bureaucratic and structured, with compliance or non-compliance a binary, absolute choice. This practice itself is enshrined within national laws and treaty obligations.

---

[7] DoD Defense Security Service, 'Comparison: Special Security Agreement and Proxy Agreement', <http://www.dss.mil/isp/foci/compar_spec_sec_prox.html>, accessed 14 October 2013.

Second, in limiting access to information, advanced states actively restrict the number of citizens entitled to knowledge of a certain subject, with implications for governance, assurance and democratic oversight – themes which are not central to this consideration. Third, by the nature of its bureaucracy and rule-bound technologies, such a control regime is expensive to operate in and enter into, making it a cost for businesses working in the defence sector. The compliance regime, therefore, could be perceived as a potential barrier to entry into the defence market for new suppliers.

The Control of Exports

Government controls over defence businesses extend beyond who can be used in government defence work and how information is to be protected. They also address the terms under which companies can move controlled technology and information beyond the borders of the state. Especially in the US, it is important to realise that information may be unclassified (and thus freely available within a country), and yet be subject to export controls because of its security implications. It can be said that there are four dimensions to the overall field of export controls.

The first dimension concerns laws and implementing regulations that address transfers beyond borders that require licences (and those that do not), including what sort of licence is appropriate, and how an application should be made.

The second dimension deals with policies on export controls, as well as government statements on which sorts of exports are likely to be allowed, which might be turned down and why. In Europe, the EU Code of Conduct on Arms Exports – since updated in 2008 to the EU Council Common Position on arms exports – is a series of policy commitments made by all EU states, listing the factors they will take into account in assessing whether or not to grant a licence.[8] An arms embargo, whether imposed on a national or multilateral basis, is an ostensibly clear form of policy.

The third dimension addresses the implementation of policy through specific and individual decisions. An example of this was Germany's decision, in February 2011, to suspend defence exports to Egypt in light of the growing opposition to former President Hosni Mubarak, while the UK and the US decided against such a blanket decision.[9]

---

[8] See 'Council Common Position 2008/944/CFSP of 8 December 2008', *Official Journal of the European Union*, L 335/99, 13 December 2008, <http://eur-lex.europa.eu/LexUriServ/LexUriServ.do?uri=OJ:L:2008:335:0099:0103:EN:PDF>, accessed 16 October 2013.

[9] Richard Norton-Taylor, 'UK Refuses to Suspend Egypt Arms Sales', *Guardian*, 8 February 2011; Brian Rohan, 'Germany Suspends Arms Exports to Egypt', *Reuters*, 7 February 2011.

The fourth and final dimension concerns the mechanisms and processes through which law and policy are implemented, and individual decisions are reached. In all cases, an application for a licence is dealt with through a consultative process among relevant government departments and by individuals who must read and draft responses to license proposals. However, there is no standardised approach on which ministry should take the lead and co-ordinate responsibility. Export licences need people to be employed to deal with applications and, unsurprisingly, the more staff employed, the quicker decisions are likely to be reached. In both the UK and the US, companies have on occasions complained about the length of time taken to reach a decision.[10]

The First Dimension: A Legal Framework

The control of exports requires a detailed set of legal arrangements and implementing regulations, and these frameworks are far from identical in Germany, the UK and the US. In part, this is due to these countries' different constitutional provisions: in Germany, people have a right to trade, whereas this is not the case in the UK.

Of the three countries under review, the US has been consistently the most concerned with the potential negative implications of technology exports for national security and, as the world's largest spender on defence research and development (R&D), it has felt that it has the most to lose. Thus it has acted as the leader in setting standards for export control which other countries have felt obliged to follow, not least to provide assurances that American technology supplied to them will be well cared for. Therefores there are some broadly common features which are identified below, before discussing the arrangements in each country in detail.

In all three countries, the fundamental drivers of whether a company should apply for an export licence are the lists of items drawn up by governments in relation to the specific goods to be covered by export-control law. As well as registers of specifically military equipment, there are directories of export-controlled dual-use goods, notably items that could support the development of chemical, biological or nuclear weapons, or ballistic missiles. In short, all three states have licensing systems which underpin their arms-control commitments under the Biological and Chemical Weapons Conventions, the Nuclear Non-Proliferation Treaty, the Waassenaar Arrangement, and the Missile Technology Control Regime. Although countries consult on the content of the lists, the three states do not control an identical set of items, particularly as regards

---

[10] See Gordon Adams, Christophe Cornu and Andrew James, 'Between Cooperation and Competition: The Transatlantic Defense Market', Chaillot Paper No 44, ISS-WEU, January 2001.

dual-use material. National lists are updated and modified on a national basis; indeed, in mid-2013, the US was engaged in reforms aimed at relaxing controls on many items by moving them from the military to the dual-use list, while moving other items in the other direction.[11] Europe, too, is considering rationalising its controls on a community-wide basis to take account of the needs of both ease of trade and security protection.[12]

Awareness of the items covered by legislation is the duty of the exporting company; as in other areas of life, ignorance of the law is no excuse and responsibility for any corporate violations of export controls lies with the chairman and chief executive. This responsibility cannot be delegated. Nevertheless, the UK government, at least, tends not to prosecute companies that make an honest mistake in not applying for a licence when it is needed.[13] The US government also tends to treat more leniently those companies that bring their export-control violations to the attention of the government, as opposed to those who hold out in the hope of not being discovered. Regardless of whether a piece of equipment or an element of information is covered by a government list, it is a company's legal duty to approach the government with a query about the need for a licence for any item the company suspects may be destined for military use.[14]

All three states have sought to control the flow of knowledge and information as well as that of products. Thus the 2008 EU Council Common Position on arms exports states that licences should be required not just for physical exports to a specified destination, but also for brokering and for 'any intangible transfers of software and technology by means such as electronic media, fax or telephone'.[15] This is a difficult area,

---

[11] C Forrester, 'US Re-Opens ITAR Category XI to Comment', *Jane's Defence Weekly*, 31 July 2013, p. 21; M Smith, 'US Announces Latest Amendments as Part of Export Control Reform', *Jane's Defence Weekly*, 17 July 2013, p. 21.

[12] European Commission, 'Strategic Export Controls: Ensuring Security and Competitiveness in a Changing World – A Report on the Public Consultation Launched under the Green Paper COM(2011) 393', EU Commission Staff Working Document, 17 January 2013; B Tigner, 'Striking a Balance: Reforming the EU's Export Control Regime', *Jane's Defence Weekly* 17 July 2013, p. 23.

[13] HM Government, *Strategic Export Control: Her Majesty's Government's Annual Report for 2010, Quarterly Reports for 2010 and 2011, Licensing Policy and Parliamentary Scrutiny Response of the Secretaries of State for Defence, Foreign and Commonwealth Affairs, International Development and Business, Innovation and Skills*, Cm 8441 (London: The Stationery Office, October 2012), pp. 14–15.

[14] See Roy Anderson, *Defence Research and Technology* (London: MoD, February 2007).

[15] See Council of the European Union, 'Acts Adopted Under Title V of the EU Treaty: Council Common Position 2008/944/CFSP of 8 December 2008 Defining Common Rules Governing Control of Exports of Military Technology and Equipment', *Official Journal of the European Union* (L335/99, 13 December 2008), Article 1.2; Keith Hartley, 'Defence Industrial Policy in a Military Alliance', *Journal of Peace Research* (Vol. 43, No. 4, 2006), pp. 473–89.

as businesspeople must consider the circumstances in which they need a licence. In many cases, it is legislation on the protection of classified information that is more important than export-control rules. Moreover, this is not a static topic, but one that is subject to change as technology evolves. For instance, the US has decided that storing information in 'the cloud' should be considered an export.

Happily for both the tourist industry and the individuals involved, the simple movement across a border of a person with controlled technology in his or her head is not considered an export. While individuals with access to information of the highest sensitivity may face limitations on their travel possibilities, this is due to the security risks involved, rather than to export-control law.

While there is significant commonality of approach between the three systems of law under consideration, each has its own particular features.

Features of German Law: Looking specifically at the case of Germany, exports of military equipment are governed by the Basic Law, the War Weapons Control Act[16] and the Foreign Trade and Payments Act,[17] in conjunction with the Foreign Trade and Payments Ordinance.[18] In addition, the Political Principles Adopted by the Government of the Federal Republic of Germany for the Export of War Weapons and Other Military Equipment of 19 January 2000 and the EU Council Common Position of 8 December 2008 provide the licensing authorities with guidelines.[19] Moreover, the Foreign Trade and Payments Act and the Foreign Trade and Payments Ordinance require the licensing of all military-equipment exports.[20]

---

[16] Act to implement Article 26 (2) of the Basic Law (War Weapons Control Act) in the version promulgated on 22 November 1990', *Federal Law Gazette I*, p. 2506 (as last amended by Article 24 of the Ordinance of 31 October 2006, *Federal Law Gazette I*, p. 2407).

[17] Revised by notification of 27 May 2009, *Federal Law Gazette I*, p. 1150.

[18] Foreign Trade and Payments Ordinance in the version promulgated on 22 November 1993', *Federal Law Gazette I*, p. 2493, last amended by the '83rd Ordinance Amending the Foreign Trade and Payments Ordinance of 19 October 2009', *Federal Law Gazette* (No. 164), p. 3737.

[19] 'Political Principles Adopted by the Government of the Federal Republic of Germany for the Export of War Weapons and Other Military Equipment', 19 January 2000.

[20] The military equipment is listed in full in Part I, Section A of the Export List (EL, Annex to Foreign Trade and Payments Ordinance). It is broken down into twenty-two positions (No. 0001 to No. 0022) that have their own sub-divisions. As with the EU's Military List, these positions are closely oriented to the corresponding list of the Wassenaar Arrangement (Munitions List), which the German government has thereby converted into national law to meet its political commitments.

By contrast, the export of military items which are not considered weapons ('other military equipment') requires – merely – a licence under the Foreign Trade and Payments Act and the Foreign Trade and Payments Ordinance.

Applicants have no legal right to the issuance of a licence for the export of war weapons. Furthermore, licences must be denied where there is a danger that war weapons will be used in connection with peace-disturbing acts or that the obligations of the Federal Republic of Germany under international law will be impaired, or where the applicant does not possess the necessary reliability. In the US and the UK, these are matters of judgement and policy, but not legal regulation. In Germany, however, things are less straightforward. Here, the federal government decides on the issuance of export licences in accordance with the discretion it must exercise under the EU Council Common Position and the aforementioned Political Principles.

The export of other military equipment is governed by the export rules in the Foreign Trade and Payments Act and the Foreign Trade and Payments Ordinance. In accordance with the principle of freedom of external economic transactions, on which the systematic approach of the Foreign Trade and Payments Act is based, the applicant has a fundamental right to the issuance of an export licence:[21]

> (1) The conducting of legal business and acts in connection with external economic transactions may be confined in order to 1. guarantee the essential security interests of the Federal Republic of Germany, 2. prevent a disturbance of the peaceful co-existence of nations or 3. prevent a major disruption of the foreign relations of the Federal Republic of Germany.

As is also the case for war weapons, the German government exercises its discretion in the issuance of export licences for other military equipment, in keeping with the EU Council Common Position and the Political Principles.

Features of UK Law: The foundations of UK legal controls is the Export Control Act 2002 which replaced the more limited act of 1939 that had been rushed through Parliament on the eve of the Second World War. In the UK, there are three main types of licence: Standard Individual Export Licences (SIELs), Open Individual Export Licences (OIELs) and Open

---

[21] Section 1 in conjunction with Section 3 of the Foreign Trade and Payments Act, unless a licence may be denied because of a violation of interests protected under Section 7 Subsection 1 of the Foreign Trade and Payments Act. See Section 7 Subsection 1 (1-3) of the Act.

General Export Licences (OGELs). SIELs 'generally allow shipments of specified items to a specified consignee up to the quantity specified by the licence', while OIELs 'are specific to an individual exporter and cover multiple shipments of specified items to specified destinations and/or in some cases, specified consignees)'.[22] OGELs 'are pre-published export licences issued by the Export Control Organisation (ECO). They are the most flexible type of strategic-export licence, as long as all pre-set conditions can be met. They allow the export of specified controlled items (identified by a "control list entry" heading) to specific destinations.'[23]

As noted, UK export controls cover information as well as goods, a reflection of the fact that government restrictions on the transfer of information affect the arms trade as well as constituting efforts to restrict the effectiveness of foreign espionage. Thus, companies are not free to move even their own privately funded and developed technological know-how if it has defence implications. In many ways, the development of these controls in the UK was partially influenced by practices in the US and, in turn, also influenced the development of the regulatory regime in Germany. Indeed, in 1980s West Germany, information to enable the licensed production of German equipment could be moved without government having a veto power, but German law has since been tightened (see Section 6 of the War Weapons Control Act referred to above).

In the UK, the need for a licence is determined by the location of the good, not its ownership. Thus a company transporting its own equipment overseas for a demonstration or exhibition must obtain a licence. A foreign company (or government) bringing controlled equipment into the UK on a temporary basis must also obtain a licence in order to 'export' the equipment back to the original state, or elsewhere.

The UK also requires a buyer to complete an end-user certificate confirming that the purchaser is the definite user of the equipment and that it is not simply to be resold to a third party. However, the UK does not insist on approving a resale that may occur after a customer has owned a piece of equipment for a significant period – in practice more than ten years.[24] In this respect, it is quite different from the US, which works hard to maintain life-long control over the destinations of its controlled technology.

---

22 Department for Business, Innovation and Skills (BIS), Strategic Export Controls, 'Country Pivot Report', 1 April 2012–30 June 2002, pp. 3, 7.

23 BIS, 'Open Licences: An Overview', <https://www.gov.uk/open-general-licences-an-overview>, accessed 16 October 2013. See also: BIS, 'Military Goods Open General Export Licences', <https://www.gov.uk/military-goods-ogels>.

24 See David Oliver, 'Current Export Policies: Trick or Treat', *Defense Horizons* (No. 6, December 2001).

British legislation includes two elements of extraterritoriality. First, it covers brokering, requiring an agent located in the UK who has arranged the movement of controlled items from one foreign country to another, without them passing through the UK, to obtain UK government approval. Second, UK citizens based overseas must also obtain British government permission should they arrange the transfer of controlled items.

Features of the US Legal System: The US government dictates extensive and rigorously enforced controls on the export of technologies, components, finished goods and services to other countries. This approach is primarily driven by a perceived need to prevent technologies and capabilities from falling into the hands of those considered 'enemies', thereby offering opponents a potential military advantage or insight into the US military's armoury. For some technologies, of course, it seems common sense for the US polity to heavily direct and regulate supply and use, including nuclear technologies or those associated with cyber-defence or offensive capabilities.

At the epicentre of the export-control process is the Arms Export Control Act, updated periodically by Congress, which controls the export of both specialist defence goods and items defined as dual-use. The boundary between the two categories is blurred and it is not unknown for items to be moved in either direction.

The US maintains lists for the two categories: the US Munitions List and the Commerce Control List. There are also two sets of detailed rules for the implementation of the legislation. The International Traffic in Arms Regulation (ITAR) covers items deemed as specialist defence goods, whereas the Export Administration Regulations (EAR) cover dual-use items.

The ITAR encompasses finished goods, components, micro-parts, training, development, advice, technical or other data and information services. Individuals can accidentally become ensnared in the ITAR directives: for example, when a non-US citizen, perhaps a research student from a US university, is exposed to a technology or item that is featured on the State Department's export-control list, the law states that the encounter must be considered an export and controlled or approved by government, theoretically in advance.[25] It is very easy to be in breach of the ITAR rules, which can attract severe federal sanctions, including unlimited fines and jail sentences. Such an environment is hardly conducive to commercial companies broadening their involvement in defence.

---

25 Directorate of Defense Trade Controls, 'International Traffic in Arms Regulation: ITAR Summary, Definitions and Subchapters', Department of State, <http://pmddtc.state.gov/regulations_laws/itar_official.html>, accessed 16 October 2013.

What distinguishes the US system is the depth of its controls, down to the technology and even knowledge level; the intensity with which they are administered; and the US's insistence on also controlling any re-export of its technologies, to the point where practices can appear nonsensical.[26] For example, if a defence business registered in the US purchases components from a supplier in the UK and these items, in the course of their useful life, need upgrading by the original UK manufacturer, the US customer would require an export licence to return them to the original provider in the UK – a licence that may or may not be granted, will add to component costs and will certainly take some time to process. This seems a bureaucratic initiative rather than a sensible security practice.

Indeed, the impact of such a culture of control on the US defence industry is profound. US companies principally focused on defence goods and services are hindered by the vagaries of the export regime, leading to regular calls for its reform. For broader commercial entities, which could potentially supply US defence contractors, the same regime can have the effect of discouraging them from accepting defence business. If a commercial technology business sells a part, either directly or indirectly, to the DoD as a component within a weapon system, that part is subject, thereafter, to export-control provisions for technology transfer. Since, proportionally, the DoD may be a much smaller customer than the global commercial market for the same technological component, this is a significant barrier to high-tech businesses becoming casual suppliers to defence contractors or the DoD.

This also has the effect of discouraging non-US customers from opting for US products or components. For example, Germany has, on occasion, urged its programme managers not to buy US goods or services because of the complex and uncertain US export-control regime and the risks this poses to its programmes.[27]

In 2013, a single line in a National Audit Office report indicated that the US may have vetoed the possibility that British F-35s might be used from a French aircraft carrier.[28] US law, policy and practice raises the sensitive question of whether countries owning US equipment should seek US permission before it is deployed on military operations or training exercises overseas. It is understood (in accordance either with ITAR arrangements or the US-UK Defence Trade Cooperation Treaty) that the US and the UK have an amendable list of approved operations and

---

[26] See Walter Pincus, 'Taking Defense's Hand out of State's Pocket', *Washington Post*, 9 July 2007.

[27] Adams, Cornu and James, "Between Cooperation and Competition'.

[28] National Audit Office, *Carrier Strike: The 2012 Reversion Decision*, Report by the Auditor and Comptroller General, HC 63 (London: The Stationery Office, May 2013), p. 11.

activities. If ever the UK wanted to use US equipment and Washington would not approve the activity, the UK would have the difficult but real choice of using the equipment anyway and dealing with the consequences later. This, incidentally, appears to be one reason why, in the event that the UK opts for a government-owned, contractor-operated (GOCO) solution for its procurement function, the government would continue to sign the relevant and necessary contracts. If a company signed, it might be held responsible by the US government for the use of the equipment.

Defence-Equipment Export Policies

As close allies with democratic political systems, considerable similarities in arms-export policy might be expected among the three countries under examination. Indeed, when the invasion of Kuwait by Iraq provoked a rethink of many Western defence-export practices, a number of common principles were agreed in resolutions of the UN Security Council and the Conference on Security and Co-operation in Europe (CSCE), now the Organization for Security and Co-operation in Europe (OSCE).

Text Box 6

The Principles Governing Arms Transfers agreed by the Forum for Security Co-operation of the Conference on Security and Co-operation in Europe

Each participating State will avoid transfers which would be likely to:

(i) be used for the violation or suppression of human rights and fundamental freedoms;
(ii) threaten the national security of other States and of territories whose external relations are the internationally acknowledged responsibility of another State;
(iii) contravene its international commitments, in particular in relation to sanctions adopted by the Security Council of the United Nations, or to decisions taken by the CSCE Council, or agreements on non-proliferation, or any other arms control and disarmament agreements;
(iv) prolong or aggravate an existing armed conflict, taking into account the legitimate requirement for self-defence;
(v) endanger peace, introduce destabilizing military capabilities into a region, or otherwise contribute to regional stability;
(vi) be diverted within the recipient country or re-exported for purposes contrary to the aims of this document;
(vii) be used for the purpose of repression;
(viii) support or encourage terrorism;
(ix) be used other than for the legitimate defence and security needs of the recipient country.

Source: Department of Trade and Investment, 'Sanction Regimes, Arms Embargoes and Restrictions on the Export of Strategic Goods: The Principles Governing Arms Transfers Agreed by the Forum for Security Cooperation of the Conference for Security and Cooperation in Europe (CSCE)', <http://bit.ly/183Q5OD>, accessed 14 October 2013.

Guidelines for Conventional Arms Transfers agreed by the Permanent Five Members of the United Nations Security Council

When considering under their national control procedures conventional arms transfers, they intend to act in accordance with the following guidelines:

1. They will consider carefully whether proposed transfers will:

a) promote the capabilities of the recipient to meet needs for legitimate self-defence;
b) serve as an appropriate and proportionate response to the security and military threats confronting the recipient country;
c) enhance the capability of the recipient to participate in regional or other collective arrangements or other measures consistent with the Charter of the United Nations or requested by the United Nations;

2. They will avoid transfers which would be likely to:

a) prolong or aggravate an existing armed conflict;
b) increase tension in a region or contribute to regional instability;
c) introduce destabilising military capabilities in a region;
d) contravene embargoes or other relevant internationally agreed restraints to which they are parties;
e) be used other than for the legitimate defence and security needs of the recipient state;
f) support or encourage international terrorism;
g) be used to interfere with the internal affairs of sovereign states;
h) seriously undermine the recipient state's economy.

Source: Department of Trade and Investment, 'Sanction Regimes, Arms Embargoes and Restrictions on the Export of Strategic Goods: Guidelines for Conventional Arms Transfers agreed by the permanent five members of the United Nations Security Council, <http://bit.ly/16ZOprF>, accessed 14 October 2013.

Within the EU, moreover, since the signing of the Code of Conduct and the Common Position, there is a continuing effort to harmonise policies in this area. In particular, the 2008 Common Position refers to eight criteria set out to guide defence-export decisions. With obvious links to the UNSC and OSCE positions, they address:

- Respect for human rights in the country of final destination
- The internal situation in the country of final destination, as a function of the existence of tensions or armed conflicts
- Preservation of regional peace, security and stability
- National security of the member states and of territories whose external relations are a responsibility of a member state, as well as that of friendly and allied countries
- The behaviour of the buyer with regard to the international community, especially regarding its attitude to terrorism, the nature of its alliances and respect for international law

- The risk that the export will be diverted within the buyer country or re-exported under undesirable conditions
- The compatibility of the export with the technical and economic capacity of the recipient.

EU states have a commitment to transparency, such that EU members have agreed to inform each other of decisions, including those not to export goods. Moreover, EU members that subsequently decide to export similar items have agreed that they must explain their decision to the other members (Article 4 of the Common Position).[29] Thus it was hoped that the Common Position would reduce the impact of the argument that 'if we don't sell it, someone else will'. Significantly, the Common Position recognises that states may feel different pressures. Although it asserts European countries' determination to set high minimum standards with regard to the control of exports and restraint, it acknowledges the wishes of member states to maintain a defence industry as part of their industrial bases and defence efforts. It also asserts that the Common Position 'shall not affect the right of Member States to operate more restrictive national policies'.

UK Policy: It has long been UK policy, albeit pursued with varying levels of zeal over the years, to promote British defence exports. This was the purpose of the governmental Defence Export Services Organisation (DESO), set up in 1964. While the Labour government under Gordon Brown (2008–10) reduced the emphasis on this point – renaming DESO the Defence & Security Organisation (DSO) and moving it out of the MoD and into the Department for Business, Innovation & Skills (BIS) – the basic recognition that the UK needed exports for its defence industry to survive did not really change.

However, it was always the case that exports were subject to controls. Successive British governments were resolutely clear about not selling when the potential purchaser was subject to a UN, NATO or EU arms embargo but, in addition, 'the longstanding British position is clear: We will not issue licences where we judge there is a clear risk that the proposed export might provoke or prolong regional or internal conflicts, or which might be used to facilitate internal

[29] Article 4 reads: 'Member states shall circulate details of applications that have been denied … together with an explanation … Before any Member State grants a licence that has been denied by another member State … for an essentially identical transaction within the last three years, it shall first consult the Member State or States which issued the denial(s). If, following consultations, the Member State nonetheless decides to grant a licence, it shall notify the Member State or States issuing the denial(s) giving details of its reasoning.'

repression'.[30] There has been considerable continuity in British governments' wish to present themselves as positive but prudent supporters of defence exports.

As with Germany and the US, British policy on defence exports also reflects the numerous international agreements to which the three states are a party and the UN Arms Trade Treaty signed in 2013 is of potentially global importance. The eight criteria in the European Common Position on defence exports rarely force a government to refuse an export. The UK has a track record of allowing, even promoting, defence exports to states that are not democracies, most notably in the Arabian Peninsula, because of British wider political and industrial interests. This approach does mean that there is occasionally some controversy about the decisions that have emerged from UK law and policy. There is a small but active group within UK society – the most prominent body being the Campaign against the Arms Trade – that opposes defence exports almost in principle and demonstrates against marketing efforts such as the Farnborough Air Show and the biennial DSEI exhibition.

German Policy: Germany has historically been more reluctant than peer states to take risks when its defence exports might lead to problems; indeed, it might be said that Germany has been wary of selling to any government that might actually need to use them. A non-legally binding political document, the Political Principles for the export of military equipment, agreed by the Social Democratic-Green coalition government under then-Chancellor Gerhard Schröder, postulated a restrictive policy on the issuing of defence-export licences.

Unlike in the US and more so than in the UK, there is an active debate in Germany on arms-export policy and its implementation. Various German commentators and opposition politicians criticise the perceived willingness of the government to export to countries with poor human-rights records, and argue that the restrictive nature of the German defence-export policy should be sustained.[31] The deputy chairman of the Social Democrats (SPD) faction in the Bundestag, Gernot Erler, has criticised the government on the grounds that 'human rights play only a subordinate role – if they play one at all – in the issuance of defence-export licences.

[30] The government position quoted in House of Commons Business, Innovation and Skills, Defence, Foreign Affairs and International Development Committee, *Scrutiny of Arms Exports (2012): UK Strategic Export Controls Annual Report 2010, Quarterly Reports for July to December 2010 and January to September 2011. The Government's Review of arms exports to the Middle East and North Africa, and Wider Arms Control Issues*, Vol. 1 (London: The Stationery Office, July 2012), p. 6.

[31] See 'Frühzeitige Veröffentlichung der Rüstungsexportberichte sicherstellen – Parlamentsrechte über Rüstungsexporte einführen', Antrag der SPD Fraktion, Drucksache 17/9188, 28 March 2012. Translation by the author.

This aspect of the political principles for defence exports, adopted in 2000, however, is a crucial component of the licence-issuance process.'[32]

Defending the government's approach to weapons exports, on the other hand, Minister of Defence Thomas de Maizière stresses that 'we [the German government] have and will maintain a restrictive defence export policy. This policy is, however, not only aligned to human rights criteria, but also to criteria relating to regional stability ... To say that human rights alone are the sole criterion is not good enough.'[33] The need to consider multiple foreign- and security-policy criteria is indeed supported by the political principles, a fact that is often (wittingly or unwittingly) overlooked by critics. The principles, as such, state that in the case of countries with questionable human-rights records, 'export licences for war weapons ... will not be granted unless in a specific case this is exceptionally warranted on particular foreign and security policy grounds, having due regard to Alliance interests'.

Given this has been common German government policy in recent years, it is not surprising that De Maizière 'cautions in particular the Social Democrats in their criticism of government policy', pointing out that the 'SPD and Greens, which are now revolting against a defence export deal, followed the very same policy when they were in government'.[34]

Germany has been, and remains, a participant in many collaborative projects which it has seen as an opportunity rather than a problem as far as exports are concerned. In principle, although details are classified, it is understood that any partner in a European collaborative project has the right to veto a proposed export. In practice, the authors know of no cases in which this has occurred although it must be expected that some potential sales campaigns were not pursued because of this consideration. Germany instead has been content to allow another partner to take the lead in terms of marketing and selling a collaborative product. This was the case prominently with the MILAN and HOT missiles that it co-developed with France as well as the Tornado. Germany would not have sold Tornado as a national product to Saudi Arabia given the latter's political system and hostility towards Israel. However, in the 1980s, BAE Systems and the UK government were allowed to pursue that market. Against the background of growing demand by so-called 'third countries', the defence-export control system is increasingly the subject of widespread controversy within Germany.

---

[32] Gernot Erler, 'Rüstungsexporte', SPD press release, 21 September 2012. Translation by the author.

[33] Thomas de Maizière, quoted in *Augengeradeaus.net*, 'De Maizière zu Rüstungsexporten', 21 September 2012. Translation by the author.

[34] Thomas de Maizière, quoted in *Spiegel Online*, 'Panzer für Saudi-Arabien: De Maizière verteidigt möglichen "Leopard"-Deal', 9 July 2011. Translation by author.

US Policy: Under the Arms Export Control Act (1976), the US government is to give consideration to whether the exports 'would contribute to an arms race, aid in the development of weapons of mass destruction, support international terrorism, increase the possibility of outbreak or escalation of conflict, or prejudice the development of bilateral or multilateral arms control or nonproliferation agreements or other arrangements.'[35] This is clearly a narrower list of concerns than that endorsed by EU states, which make reference to human-rights abuses and the capacity of a state to afford equipment, but, as seen with the OSCE and UNSC provisions, the US has since recognised a longer list of relevant concerns.

Until 2007, the US had shown special treatment – in the form of slightly relaxed controls – only to Canada. However, that changed when President George W Bush agreed with Prime Minister Tony Blair on a UK-US Defence Trade Cooperation Treaty under which a specified list of items would not need licences for export to the UK, provided they were being sold to a company on an approved list. The same arrangements would apply to exempted items going from the UK to the US. The treaty was not ratified by the US until 2010, with the modest list of exemptions including much unclassified (but still controlled) material.[36] As of mid-2013, about a dozen companies had applied and been accepted on to the approved list, with about three times that number considering the matter. What was unclear was how many of the approved companies had sought that status because they felt it was diplomatic in light of all the effort officials had made to get the treaty in place, or whether they expected real advantages and savings. Australia is the only other country to have a similar treaty with the US, but neither Australia, nor Canada nor the UK has any special discretion about the re-export of US goods.

Policy Implementation and Export Controls

Wider appreciation of similarities and differences among Germany, the UK and the US must take account of what they do as well as what they say regarding export controls – that is, how the regulations and policy are actually used in specific circumstances.

German Policy in Practice: The key ticket items in Germany's 2010 defence-exports efforts (issued licences and actual exports) were high-value armaments like submarines, warships and tanks. Some 71 per cent of the individual export licences related to exports were designated for EU, NATO and NATO-equivalent countries, and 29 per cent for other

35 US Congress, Arms Export Control Act 1976.

36 See B Salzman, 'Passed At Last: The US/UK Defence Trade Treaty', 7 October 2010.

countries including India, Pakistan, Saudi Arabia, Singapore and the UAE. In terms of value, 77 per cent of all weapon exports in 2010 were designated for EU, NATO and NATO-equivalent countries, and 23 per cent were delivered to third countries. Developing countries received German weapons with a total value of €108.5 million (5.1 per cent of all German weapon exports in 2010, compared to 3.9 per cent in 2009).

Moreover, in 2013 the defence-export agenda demonstrated the pressures on the government to interpret its policy in a more relaxed manner, especially with regard to non-democratic and newly democratic governments, and the political disagreements this pressure generated.

In July 2012, the German government confirmed that the government of Qatar had expressed interest in buying up to 200 Leopard 2 main battle tanks (MBTs) built by Munich-based arms manufacturer Kraus-Maffei Wegmann (KMW). Thus, Qatar joined a highly controversial group of potential buyers of German-made, high-tech military kit, including Saudi Arabia, Indonesia and Egypt.

Both Saudi Arabia and Indonesia were also interested in acquiring the Leopard 2: 270 of the most advanced A7+ version in the case of the former and 130 of the second-hand A4 for the latter. Egypt, for its part, was seeking to buy two *209*-class submarines built by Kiel-based dockyard HDW, which had already sold three of the same to Israel, and was due to deliver a further three by 2017. Algeria had signed a €400 million contract with German company ThyssenKrupp Marine Systems for the delivery of two *MEKO A-200*-class frigates, including on-deck helicopters, by 2017.

Although the requests from Doha, Riyadh, Jakarta and Cairo had not yet been approved by Germany's Federal Security Council (BSR) in mid-2013, this list of high-value defence exports for 'third countries' raised substantial concerns. However, the utility of MBTs such as the Leopard 2, as well as submarines and frigates, as instruments of internal repression was questionable, which mitigated human-rights sensitivities. De Maizière also highlighted this point, emphasising that 'the type of defence export is a further criterion for the issuance of an export licence'.[37]

It is, at present, too early to make a final judgement on the government's compliance with the Political Principles. On the one hand, at the 2012 Bundeswehr convention – an annual gathering of senior command personnel of the Bundeswehr and the BMVg – Angela Merkel reiterated that the government 'does not intend to dilute the restrictive defence export policy or to weaken the importance of human rights'. On the other hand, she also stated that: 'if you feel an obligation to undertake peacekeeping but cannot take on an active peacekeeping role everywhere in the world, you are also called upon to support reliable partners in their

[37] Thomas de Maizière, quoted in *Die Welt*, 'Verteidigungsminister de Maizière verteidigt Rüstungsexporte', 5 August 2012. Translation by the author.

efforts to undertake such tasks', arguing further that: 'other nations and regional organisations should not only be encouraged, but should also be enabled to solve specific conflicts themselves through training and equipment support'.[38]

As such, Merkel echoes one of her previous statements from October 2011, in which she explained that: 'it is generally not enough to send other countries and organisations words of encouragement if Germany shies away from military intervention', instead suggesting that Germany must also provide the necessary means to those nations that are prepared to get involved, emphasising that this includes arms exports.

The repeated outlining of such an approach – which bears a certain resemblance to the third major tenet of the 1969 Nixon Doctrine, which postulated that the US would provide military and economic assistance when requested in accordance with treaty commitments, but would look to the requesting nation 'to assume the primary responsibility of providing the manpower for its own defence' – could be seen as an early sign of a qualitative change in Germany's defence-export policy, and one that would eventually require a revision of the Political Principles.

UK Policy in Practice: There is significant transparency about the number and nature of licences being approved or rejected in the form of a lengthy annual report published by the Foreign Office, a report which is regularly scrutinised by a special grouping from the House of Commons which brings together members of the Defence, Foreign Affairs, International Development and BIS Select Committees.[39] Indeed, for Britain, its 'Strategic Export Controls Reports demonstrate the Government's commitment to transparent and responsible controls on British arms exports. It gives details by destination, both of physical exports and of what equipment has been licensed for export.'[40]

However, the implementation of export controls in the UK has not escaped controversy, with several areas prompting periodic or even regular attention from the press and Parliament. There have been a number of cases, of which the most prominent involved the Al-Yamamah deal with Saudi Arabia, which linked defence businesses and even the government with the use of bribes. As a result, BAE Systems introduced a code devised by Lord Woolf, a senior judge, to guide the behaviour of BAE Systems staff. In general, the defence industrial sector is often suspected of

[38] Angela Merkel, quoted in *N-TV.de*, 'Instrument der Friedenssicherung: Merkel wirbt für Rüstungsexporte', 22 October 2012. Translation by the author.

[39] HM Government, *United Kingdom Strategic Export Controls Annual Report 2011* (London: The Stationery Office, July 2012).

[40] UK Foreign Office homepage, <http://www.fco.gov.uk>, accessed 12 June 2013.

particular involvement in corrupt practices and receives considerable attention from Transparency International, the prominent anti-corruption non-governmental organisation.[41] In this respect, a key British consideration is that defence companies are not granted any special exemption from wider law, particularly from the Bribery Act which came into force in 2012. The events of the Arab Awakening, since December 2010, particularly highlighted possible risks to the UK's long-term interests of supporting – through arms supplies – regimes that could fail in the near future. Successor governments in the Middle East could be actively hostile to British defence and economic interests because of the UK's past readiness to work closely with and support the previous regime. Events in Egypt since 2011 have highlighted these particular challenges in terms of policy implementation – not just for the UK, but for all Western governments providing some form of assistance to the Egyptian military, although the dilemma loomed largest for the US as Egypt's largest supplier in this regard. Significantly, and probably predictably, the UK avoided defining the overthrow of President Morsi in July 2013 as a military coup, instead calling it a 'military intervention'. This had the effect of making it easier for the US government to take a similar stance, meaning that it could avoid the requirement, under US law, to suspend military aid to Egypt in the wake of a military coup.[42]

US Policy in Practice: For the US, defence exports are a key element of its foreign and security policies; its military aid programmes, including the Foreign Military Financing funds covering grants for US-provided military goods and services, are far larger than those of the UK and Germany.[43] As the reluctance in the summer of 2013 to place an embargo on Egypt showed, and as the response to Israeli uses of US-origin equipment over decades has indicated, Washington does not easily allow its long-term concerns to be blown off course by what can be viewed as temporary setbacks.

---

[41] See Transparency International, 'Our Work With the Defence Industry', <http://www.ti-defence.org/what-we-do/defence-industry>, accessed 17 October 2013; in particular, Martin Pyman et al, 'Defence Companies Anti-Corruption Index'.

[42] H Alexander, 'World Reaction to Egypt Coup', *Daily Telegraph*, 4 July 2013; Robert Fisk, 'When is a Military Coup not a Military Coup', *Independent*, 4 July 2013; K Ghattas, 'Egypt is Still not a Military Coup in Washington', *BBC News*, 18 July 2013.

[43] See US Department of State, 'Foreign Military Financing (FMF)', <http://www.state.gov/t/pm/65531.htm>, accessed 2 October 2013; see also 'Foreign Military Financing Account Summary', <http://www.state.gov/t/pm/ppa/sat/c14560.htm>, accessed 2 October 2013; US Census website, <http://www.census.gov/compendia/statab/2012/tables/12s1299.pdf> and <http://www.census.gov/compendia/statab/2012/tables/12s1298.pdf>, accessed 16 October 2013.

As noted, when a component is embedded within a military system, or is deemed to possess a military utility, it becomes subject to export controls under the ITAR and the Export Administration Regulations (EAR). Off-shore sales of that item, or the system within which it is used, are therefore also subject to US government approval. This means, of course, that the US government controls, on a life-long basis, the world-wide supply of technology developed and manufactured from within its borders, irrespective of whether the item is principally intended for civil or military use. In practice, therefore, US export controls apply universally to all goods and services originating, in whole or in part, from within the fifty states.

The export regulations in the US make a clear distinction between direct commercial sales from a US firm to a foreign government of which the US authorities approve, and Foreign Military Sales which are undertaken by the US on a government-to-government basis. Examples of the latter include the $20-billion land- and air-systems export package from the US to Saudi Arabia in 2007, and a further $20-billion package to Bahrain, Kuwait, Oman, Qatar and the UAE in the margins of the same deal.[44] In practice, this deal was driven by political considerations; a reward to Gulf Cooperation Council states for their support to the so-called war on terror whilst, simultaneously, locking in key countries as dependants of the US. As is demonstrated through this example, export controls are political rather than commercial or bureaucratic tools, concerned with promoting and protecting US power.[45]

Such large contracts as these suggest that the processes for foreign sales of defence equipment and the corresponding transfer of technologies from the US to its allies comprise a tried and tested journey for suppliers, purchasers and regulators. This is far from the case. In reality, there can be significant systemic delays associated with the securing of export licences from US authorities, mostly because of the vast bureaucracy and administrative distribution of powers and responsibilities across the federal system. Each year there are roughly 80,000 applications for export approvals that have to transit first through the State Department and then, if endorsed, through to the Pentagon before an export licence can be granted. Invariably, there is an eight-to-ten week waiting list for papers even to be entered into the system. Delays of up to a year for the entire process are not exceptional. It is therefore possible to deduce that the US export regime is both highly political and highly bureaucratic; in fact this very bureaucracy can often help in the protection of US

[44] Simeon Kerr, 'Oil Rich States Step Up Market Presence', *Financial Times*, 10 September 2007.

[45] Jacques S Gansler, *Democracy's Arsenal: Creating a Twenty-First Century Defence Industry* (Cambridge, MA: MIT Press, 2011), p. 151.

technologies, as tired and frustrated commercial applicants give up engaging with the system.[46]

Among many industry and government officials in Europe there is a sense that the US on occasion uses its export controls for commercial purposes: in other words, to hinder the sale of larger European systems that incorporate US technology. One possible example of this was the inability of Raytheon UK, between 2011 and 2013, to get permission for the export of its Paveway IV laser-guided bomb to Saudi Arabia (most of the bomb is made in the UK but some parts come from the US). One of the explanations mooted was that industrial lobbyists in the US were concerned that the integration of Paveway IV into Saudi Typhoons would make the aircraft a more challenging competitor for its US counterparts in the global marketplace. Reports in July 2013 that the US was about to change its position were then thought by some to be an attempt to prevent an alternative weapon, the French Armement Air-Sol Modulaire, from being integrated with the Saudi Typhoons, having already been fitted to the country's Tornado aircraft. The formal US position was that it did not want the precision-bombing capability of the Paveway IV to be made available in the Middle East.

The US Arms Control Association is among those that believe that American restrictions are applied less rigorously when it comes to relatively non-sensitive technologies including small arms.[47] This sense of a hierarchy of controls in practice is beginning to cause concern both in Washington and the governments of allies that arms controls are pliable in practice and less rigorous then their advocates would propose.[48]

Governmental Decision-Making

The different governments under analysis in this monograph obviously have different ministerial structures, but they also have different views as to which ministry should lead on export-control issues.

In Germany, the BMWi is the top authority for the protection of classified information by industry, as well as being the licensing agency for commercial transactions.[49] The other ministries whose portfolios include the treatment of weapons of war (finance, interior and defence) are

---

[46] Lincoln Bloomfield, Jr, *Export Controls and Technology Transfers: Turning Obstacles into Opportunities* (Washington, DC: Hudson Institute, 11 September 2006).

[47] B Benowitz and B Kellman, 'Rethink Plans to Loosen US Controls on Arms Exports', *Arms Control Today* (April 2013).

[48] See Gansler, *Democracy's Arsenal.*

[49] 'Gesetzes über die Voraussetzungen und das Verfahren von Sicherheitsüberprüfungen des Bundes (Sicherheitsüberprüfungsgesetz – SÜG)', para. 25.

themselves responsible for the respective approvals falling within their scope of competence.

The Federal Office of Economics and Export Control (BAFA), which is a subordinate agency operating under the jurisdiction of the BMWi, is responsible for granting and denying export licences under the Foreign Trade and Payments Act and Ordinance.[12] Any projects that may have particular political impact are submitted by BAFA to the federal government for its political assessment. Germany's Federal Security Council – a Cabinet committee chaired by the federal chancellor – is normally included in deliberations on export projects that stand out because of the consignee country, the military equipment involved, or the volume of the transaction. The committee's members comprise the ministers of foreign affairs, finance, interior, justice, defence, economics and technology, and economic co-operation and development. The handling of export-control issues by the German government therefore appears highly bureaucratic, poorly co-ordinated, contradictory and confusing, despite the existence of the various related documents, practices and guidelines.

In the UK, even for defence products, the licensing issuing department is BIS, although there is extensive consultation with other concerned departments, especially the Ministry of Defence and the Foreign Office. The intelligence sector also has a voice where relevant.

Both the UK and Germany have processes for dealing with problems posed by the fact that neither potential purchasing governments nor selling companies want to engage in extensive negotiations without knowing the likelihood of an export licence being granted. In Germany, the so-called advance inquiry practice has become customary during the past several decades. This practice enables companies to clarify at an early stage whether, upon agreement of a sales contract, the required export licence would be granted at a later point in time – assuming that the circumstances of the transaction remain unchanged. Decisions on these advance inquiries are taken in accordance with the same criteria as decisions on export-licence applications.

Advance inquiries relating to weapons of war must be submitted to the Federal Foreign Office, while those relating to other military equipment must be filed with the Federal Office of Economics and Export Control – as is the case with actual licence applications. In this situation, important projects are submitted to the federal government for decision. The purpose of advance inquiries is to make the outcome of the subsequent licensing procedure visible at the earliest possible stage in the interest of reliable planning, although it is not a substitute for the actual export licence.

Table 10: Examples of Trade Control Licences.

	France	Germany	Japan	USA	Indonesia	Global Average
SIELs/SITCLs						
Median processing time (days)	13	12	6	12	17	14
No. completed in 20 working days	79% (99)	83% (105)	91% (11)	82% (209)	71% (33)	17 (2,958)
No. completed in 60 working days	97% (121)	100 (126)	91% (11)	98% (248)	95% (44)	95% (3,905)
Cases refused	0%	0%	0%	0%	0%	1% (49)
OIELs/OITCLs						
Median processing time (days)	33	23	134	76	87	61
No. completed in 20 working days	44% (11)	47% (11)	14% (1)	22% (6)	14% (1)	26% (35)
No. completed in 60 working days	68% (17)	56% (13)	14% (1)	44% (12)	42% (3)	49% (62)
Cases refused	0%	0%	0%	0%	0%	2% (30)

Source: Department for Business, Innovation and Skills.

In the UK, some de facto guidance can be obtained through the MoD Form 680 (F680) process. In simple terms, the marketing of a potential export usually requires that the prospective customer be provided with information about the system's characteristics and performance, and some of that information is normally classified. To be able to release such information, companies must make an application on a specific form to the relevant authority – in this case, the MoD, which is deemed to have the lead on assessing whether UK military advantage would be put at risk. According to a government website, 'for companies about to embark on a marketing campaign the F680 gives an indication of whether an export licence would be likely to be approved, but does not remove the need for an export licence'.[50]

At a very practical level, in the past there has been overt concern in the UK about the time taken to secure licences because of the volume of work involved and the limited number of staff in BIS and elsewhere devoted to the task of handling applications. Significant effort has been dedicated to this area and the British government publishes quarterly information on this topic. Table 10 comprises extracts from the BIS Quarterly Report (covering April–June 2012) and shows, predictably, that SIELs and Single Individual Trade Control Licences (SITCLs) are processed

[50] See UK Trade and Industry, 'Export Control', <http://www.ukti.gov.uk/defencesecurity/defence/exportcontrol.html>, accessed 2 October 2013.

more quickly than OIELs ('SIELs generally allow shipments of specified items to a specified consignee up to the quantity specified by the licence … OIELs are specific to an individual exporter and cover multiple shipments of specified items to specified destinations and/or in some cases, specified consignees').[51] Incorporating a sample of EU, NATO, friendly and major developing states, the table indicates that OIELs in particular are processed more quickly for major EU destinations than for the US or Japan.

In the US, there is perhaps more uncertainty in a commercial sale, although clearly a company must get permission to release US classified information about a system to a foreign government. However, if a country seeks to acquire a system through the FMS programme, most of its dealings throughout the process will be with the US government. An additional feature of the US is that all prospective exports with a value in excess of $50 million must be submitted to Congress which then has thirty days to veto it. The absence of any explicit vote against a sale in the time allowed is deemed to constitute consent, but this means that US officials are in regular contact with Congress about possible exports, even under the FMS programme.

The biggest difference with regard to governmental decision-making in this area between the US, Germany and the UK is that the lead ministry with responsibility for co-ordinating the views of others in the latter two states is the trade and industry ministry, whose emphasis will naturally be on developing the national economy through sales. In the US, by contrast, the State Department leads, with an emphasis on national security, and is responsible for the issuance of export licences for items on the Munitions List. The Department of Commerce leads only on dual-use technology. The State Department is unlikely to give this power up lightly, despite its apparent unsuitability to this role, given that in many other areas of national security and defence policy it can easily be overshadowed by the DoD.

Implications of Governmental Decision-Making on Defence Businesses: The controls on defence exports constitute an administrative and financial burden on defence companies, which must employ significant numbers of trained people to complete licence applications and ensure company compliance. When components and sub-systems associated with collaborative projects such as Typhoon and the A400M are moved from one state to another, they still require export licences, although open licences are available. This represents a distinct burden for multinational defence businesses such as Thales and Raytheon, which may wish to manufacture sub-systems in one country and move them to a second state for incorporation into a larger product. It is unsurprising, therefore, that the

[51] BIS, Strategic Export Controls, 'Country Pivot Report', pp. 3, 7.

European Commission has sought to harmonise and ease the rules for the transfer of defence products within the EU.[52] However, it has had to recognise the ultimate requirement for national direction, not least with regard to policy.

The demands of US law and practice are particularly onerous in this regard, given the insistence on approving all further transfers of controlled goods.[53] Thus a British aircraft (with 10,000 components) which might contain twenty sub-sub-systems of US origin will need twenty licences from the US before it can be sold to a third party. If the aircraft is returned to the UK for a major overhaul, it will need a further round of licences, and yet another before it can be returned to its 'owner'. Not surprisingly, offering a defence product as 'ITAR-free' is often viewed as a positive sales feature by companies outside the US.

Rather hidden beneath this obvious challenge for the defence industry and government is the information-management issue, which requires governments and companies to keep track over decades (and throughout product upgrades) of the licensed technology present in a large system.

Defence Companies, Cartels and Non-National Investments

The domination of a sector of the economy by a small number of firms or a single company opens up the possibility of collusion between and price fixing by supposedly competing firms. Thus in the UK, Germany and the US there is anti-trust legislation targeting cartels.

The European Commission has acted as the main regulatory body of the civil economy in Europe since the establishment of the single market.[54] However, the Commission has no powers regarding specialist

---

[52] EU Commission, 'Directive 2009/43. EC of the European parliament and the Council, 6 May 2009: Simplifying Terms and Conditions of Transfers of Defence-Related Goods within the Community', *Official Journal of the European Union* (10 June 2009).

[53] See, for instance, 'Why Do US Export Controls Affect Non-US Companies', <http://www.thefreelibrary.com/US+Export+Controls+and+Non-US+Companies-a01073983734>, accessed 2 October 2013.

[54] Any merger projects with a Community dimension are examined by the European Commission in Brussels. According to article 1 of 'Council Regulation (EC) No 139/2004 of 20 January 2004 on the control of concentrations between undertakings (the EC Merger Regulation)', a concentration has a Community dimension where '(a) the combined aggregate worldwide turnover of all the undertakings concerned is more than EUR 5000 million; and (b) the aggregate Community-wide turnover of each of at least two of the undertakings concerned is more than EUR 250 million, unless each of the undertakings concerned achieves more than two-thirds of its aggregate Community-wide turnover within one and the same Member State'.

In addition to these thresholds, EC No 139/2004 assumes a concentration with a Community dimension where '(a) the combined aggregate worldwide turnover of

defence firms, whose products are largely covered by Article 346 of the Lisbon Treaty. For example, BAE Systems' rise to dominance in the UK military aircraft, shipbuilding and submarine sectors came through mergers allowed by the UK Ministry of Defence but in which the EU had no voice. Had the BAE Systems merger with EADS moved forward in October 2012, the Commission's role would have been an interesting case study, as BAE Systems is a predominantly – but not exclusively – defence-sector company.

At the national level, governments may – and do – treat defence companies differently, as the scale of national markets and the high costs of maintaining expertise in different defence sectors have forced governments to embrace monopolies and duopolies in different areas. Across Europe, in sectors including naval shipbuilding and tracked armoured vehicles, the problem is that there are still too many suppliers (unlike in the US, where there are often too few). However, the British Ministry of Defence still had to secure special government dispensation for the creation of Team Complex Weapons, which provides a set of companies, such as MBDA and QinetiQ, with an assured range of work so that the UK more broadly can maintain a capability to design and develop, build, test and support a range of modern missiles. This approach thus entailed anti-competitive behaviour.

An example of national cartel regulation in the German defence sector was the reintegration of army maintenance, which had previously been privatised through a PPP agreement which created the company Heeresinstandsetzunglogistik (HIL). The Federal Cartel Office initiated a procedure to reverse the privatisation due to potentially conflicting interests on the industry side – with the company acting both as contracting authority and contractor – and, ultimately, it prompted the BMVg to repurchase shares of the company held by the private sector, thereby changing HIL from a PPP to an in-house company.[55]

By contrast, the US government famously encouraged the consolidation of defence prime contractors in the US after 1993 with its 'Last Supper'[56] and then halted the process when it prevented the merger of

all the undertakings concerned is more than EUR 2500 million; (b) in each of at least three Member States, the combined aggregate turnover of all the undertakings concerned is more than EUR 100 million; (c) in each of at least three Member States included for the purpose of point (b), the aggregate turnover of each of at least two of the undertakings concerned is more than EUR 25 million; and (d) the aggregate Community-wide turnover of each of at least two of the undertakings concerned is more than EUR 100 million, unless each of the undertakings concerned achieves more than two-thirds of its aggregate Community-wide turnover within one and the same Member State.'

[55] Bundesregierung, 'Regierungspressekonferenz vom 15. August 2012', <http://www.bundesregierung.de/Content/DE/Mitschrift/Pressekonferenzen/2012/08/2012-08-15-regpk.html>, accessed 2 October 2013.

Northrop Grumman with Lockheed Martin in 1999. The US has a legally based body, the Committee on Foreign Investment in the US (CIFIUS), with the power to approve or reject any major foreign investment – which can significantly affect the degree of concentration in the sector – including in the media and entertainment industry. Given the country's firm control of classified information (as discussed earlier in this chapter), Washington has generally been ready to see extensive foreign investment in the American defence sector, especially by British-based companies, not least because it increases competition. Of the major interested defence firms, only Thales has found it hard to complete investments in the US, with Finmeccanica having bought DRS Technologies at the beginning of the century.

Germany, on the other hand, whose government has the power to reject any acquisition of a German defence company, appears to have less confidence in its controls on information and more concerns about future investment and commitment, and so has been reluctant to allow foreign investment in its defence sector. For instance, it would not allow Thales to buy STN Atlas and in 2012 prevented EADS from merging with BAE Systems.

As far as the UK is concerned, consolidation issues related to the defence industry have largely been handled on a case-by-case basis, although the UK government does hold a 'golden share' in both Rolls-Royce and BAE Systems, which gives it power over whether the companies can be bought and by whom. Britain has been much more open than Germany to foreign defence industrial investment, with the consequence that many of the MoD's major sources of supply are foreign-owned, although substantial elements of the development and production work are undertaken in the UK.

In terms of UK employment numbers, the country's largest defence firms after BAE Systems are Finmeccanica (which owns AgustaWestland and Selex) and Thales. Lockheed Martin, Raytheon and Northrop Grumman all add sufficient value in the UK to count as 'British' firms under the criteria set out in the 2002 Defence Industrial Policy paper, as does EADS/Airbus. In truth, the British government had little choice but to allow foreign investment in the defence sector since, under the pressures of the competitive tendering policies of the 1980s and 1990s, and with the limited defence budgets of the post-Cold War era, many defence firms, including General Electric Company, Plessey Semiconductors, Racal Electronics and

[56] In July 1993, US Secretary of Defense William Perry urged the top US defence companies to merge and consolidate to reduce overhead as procurement declined. At a Pentagon dinner (dubbed the 'Last Supper'), Perry pledged to support efforts by the defence industry to undertake its own consolidation.

Shorts Missile Systems, put themselves up for sale and there was no British buyer available except, possibly, for BAE Systems.

Restriction on Career Transfers

In addition to their ability to recruit staff that have not got or cannot obtain security clearance, defence companies also have restrictions about their ability to hire individuals who *do* have such clearances.

Governments often wish to control the ability of their own defence staff to move from government employment to the private sector, usually for two reasons. The first is that such staff could take with them specialist insider information which could benefit a particular company in a contract competition. The other is to reduce the temptation on the part of government defence staff to favour a particular company during the tendering process in the expectation that they could seek employment there having left government service.

In Britain, the formal rules are that military and civil-service staff must obtain permission to work in a defence-related business, although awareness of and respect for these rules is not universal. With competitors of a prospective employer also having a voice, the norm is that government employees are expected to leave a gap of between three and twelve months before going to work in a private-sector business related to their previous employment. Anecdotal evidence suggests that senior personnel feel the pressure of these rules rather more strongly than middle-ranking staff. Restrictions, however, must take account of the reality that the UK – like Germany – as a signatory to the European Convention on Human Rights must recognise that individuals have a basic right to work for the employer of their choice.

While in the UK the 'revolving-door' issue is a matter of periodic concern, in Germany there are firm government measures in place. The private sector has a natural interest in recruiting former soldiers and civil servants of the Bundeswehr in order to acquire their substantial experience in the public domain, their specific user expertise and their understanding of operational requirements, and through which to exploit their ability to address issues in a language known to the customers and to gain access to established public-sector networks. However, the legal restrictions regarding career transfers of all retired soldiers and former civil servants – not just those at the more senior levels – from the public to the private sector, as well as their secrecy obligations having concluded service, are particularly rigorous in Germany.

According to German law, every retired career soldier or former soldier with service-pension entitlements must report to the BMVg or a subordinate body a new occupation or any other employment outside the public sector that is related to his official activities within the five years prior to retiring from the service and may impair official interests – and they must do so before commencing it. If the BMVg finds that official interests would be impaired, it may forbid the individual from taking up this employment for the maximum period of the disclosure obligation (five years after leaving the service), unless there are grounds for a shorter prohibition period. Civil servants of the German armed forces are subject to the same restrictions. Another way of tackling this issue in Germany is the generally longer time in service for military personnel than in the UK, meaning that many former soldiers, in receipt of a pension, feel less inclined or have less time in which to undertake a second career in the private sector. The normal retirement age for a UK one-star officer is fifty-five, with more senior officers retiring at the age of fifty-eight or younger. German officers of the same rank can serve in uniform until they are sixty-five years old.

The BMVg's Internal Affairs Department (Referat Ermittlung in Sonderfällen – ES) has more or less consistently applied this legal framework, and especially in high-profile cases involving senior officers that have become a topic of public discussion. As an illustration, consider the case of retired Lieutenant General Heinz Marzi, who was forced by the ES department to step down as managing director of the Federal Association of the German Security and Defence Industries (BDSV) in November 2010 – eleven months after he undertook the role and twenty months after he had retired from the Bundeswehr – due to conflicting interests with his former official position.[57] His successor, Georg Wilhelm Adamowitsch, a former permanent state secretary of the BMWi, was not subject to the disclosure obligation when he took over as managing director in 2011 because he did so more than five years after he had retired as a civil servant.

In addition to the disclosure-obligation period, former soldiers and civil servants of the Bundeswehr are obligated to secrecy after leaving on all classified official matters with which they became familiar during their official activities. Notwithstanding the restrictive nature of this legal framework, many former soldiers and civil servants still find their way into the private sector in the defence industry, either because they wait until the disclosure-obligation period has passed or because the ES department has no reasonable grounds on which to prohibit a career transfer. Moreover, the utilisation of non-classified knowledge – for example, regarding structures and processes relevant to acquisition – and the application of specific competencies – like the command of technical and

[57] *Spiegel Online*, 'Rüstungsindustrie: Verteidigungsministerium stoppt Lobbyarbeit von Ex-General', 6 November 2010, <http://www.spiegel.de/politik/deutschland/ruestungsindustrie-verteidigungsministerium-stoppt-lobbyarbeit-von-ex-general-a-727674.html>, accessed 2 October 2013.

operative language – are highly valuable assets for corporate business development and strategic orientation in the long term.

In the US, the regime is more relaxed, not least because there is a tradition of people being brought in from the private sector to work in government on a temporary basis, and of retired government employees going to work in consultancy and elsewhere in the private sector.

Text Box 7: UK Search-and-Rescue Helicopter Services.

In 2011, the consortium organisation Soteria, the designated preferred bidder for a UK government contract to provide a helicopter-based search-and-rescue service, admitted it had enjoyed access to commercially sensitive information. It was alleged to have hired a military officer from the government's bid team.

As information came to light, and after the Royal Bank of Scotland had withdrawn from the Soteria consortium, the government was compelled to halt the competitive process. When it was later re-started, Soteria opted not to take part and the contract was eventually awarded in 2013 to the Bristow Group, which would operate a mixed fleet of Sikorsky and AgustaWestland helicopters.

Anti-Bribery and Anti-Corruption Measures

Limits on the ability of government employees to move to the public sector are clearly linked to concerns about bribery and corruption. Anti-bribery measures are an important component of the regulation of corporate behaviour in defence – as in all other commercial sectors. Corruption is a potential issue in both the domestic and international defence markets.

Defence-procurement scandals are considered by some to be part of the fabric of any advanced industrial society whose government has the ambition to play a role on the international stage. The United States is no exception. In the mid-1980s, a survey was commissioned by the DoD which revealed that the population of the US considered waste, fraud and general abuse within the defence-acquisition system to be a national disgrace, and solving it a cross-party political imperative.[58] Indeed, respondents believed that close to half of the defence budget was lost to some kind of waste or fraud and that contractors, invariably, were dishonest. There is little data to support this popular narrative which, somehow, endures, but the key point is that the US government is sensitive to it and strives to demonstrate that it is actively engaged in interdicting wasteful, fraudulent and abusive practices. However, it is important to define the relevant terms, as they cannot be used

[58] Packard Commission, 'Survey of Public Opinion on Defense Procurement', GPO, 1986.

interchangeably:[59] waste is the inefficient and ineffective use of government money through poor management; fraud is the perpetration of an illegal act; and abuse is an unintelligent wrong, but not strictly illegal.

The Grace Commission – an investigation into waste and inefficiency in the US government initiated by President Reagan in 1982, formally known as the Private Sector Survey on Cost Control – identified over 100 categories of waste in the DoD, with progress in managing and eliminating these items to be reported by the department at least annually to Congress.[60] The three leading categories, by occurrence and value, were high overheads charged by contractors to government, instability in budgeting and a lack of cost awareness in the design phase of defence programmes. President Obama's Better Buying Power (BBP) initiative – launched in 2010 – and its derivative BBP 2.0, is the ongoing government programme with the aim of directing federal responses to these perceived challenges.[61]

However, the reality of waste and inefficiencies is different from the dominant narrative of management failure and corruption. Whilst the US endured a number of high-profile procurement scandals in the 1980s and 90s, it was clear from the DoD's inspector general's analysis that 'for every dollar wasted ... only two cents are stolen; the rest is lost because of mismanagement.'[62]

Indeed, the sense that fraud is widespread is misplaced. One analysis concluded that of 330,000 US acquisition programmes reviewed only 372 were considered problematic, a statistically insignificant value.[63] Another source concurs that fewer than 0.1 per cent of the 15 million annual acquisition purchases in the US involve some kind of fraud or abuse.[64] The data therefore indicate that the challenges for US defence acquisition are not to be found in incidences of fraud or malpractice.

The same is true of the UK and Germany: corrupt practices in domestic defence procurement cannot be entirely eliminated but they are not a prominent issue. Transparency International's Defence Index lists Germany as being in the best category in terms of the rarity of corrupt practices, while the US and the UK are in category B, which locates them

---

[59] Gansler, *Democracy's Arsenal*, p. 193.
[60] See the Grace Commission, 'Final Report of the President's Private Sector Commission on Government Management', GPO, 1985.
[61] See Center for Strategic and International Studies (CSIS), *Better Buying Power Initiative 2.0* (Washington, DC: CSIS, 14 November 2012).
[62] P Earley, 'Sherick Seeks to Plug Pentagon Dyke', *Washington Post*, 26 November 1984.
[63] Gansler, *Democracy's Arsenal*, p. 197.
[64] *Ibid.*

all in the best nine states in the world.[65] However, this is in part because of government control measures, some of which may have the adverse consequence of hindering effective management. For instance, UK project team leaders' time in post is deliberately restricted (five years is exceptional) so as to prevent the development of overly intimate relations with a prime contractor. But that can induce a readiness to take high-risk decisions, not least because the post-holder wants to make his or her mark on a project in the time allotted: by the time the consequences become apparent, the project leader will be in a different post and perhaps even have been forgotten. Companies, of course, must be aware of the somewhat different 'marginally acceptable' practices in different countries, as exemplified by the varying attitudes to providing corporate entertainment to customers.

In Germany, although the government is not involved in the criminal prosecution of acts of bribery, it can exclude companies and individuals who have been found guilty of such crimes (and whose reliability is therefore questionable) from public tenders and refuse to issue them with a licence for the export of products. This is also true in principle in the UK and the US. However, when corruption does occur, in all three states the monopolistic and oligopolistic nature of the defence market makes it impractical to ban a company convicted of corrupt practices for any period of time: for example, in 2005, very senior Boeing officials were jailed over the Darleen Druyun affair, but Boeing were not placed on what was then known as the Excluded Parties List System (and so were not ineligible for government or even DoD contracts) for even a brief period.[66] BAE Systems similarly paid significant fines ($400 million) in the US in settlement of allegations relating to the Al-Yamamah deal and payments into a US bank, but it remained an eligible US contractor.[67]

Defence exporting, however, is widely recognised as vulnerable to corrupt practices and is thus the focus of particular attention by Transparency International.[68] The basic relevant consideration is that all three countries under scrutiny in this monograph are signatories to the Organisation for Economic Co-operation and Development (OECD) anti-bribery convention and have legislation in place that builds these commitments into domestic law. Moreover, under national laws, all three

[65] Germany sits alongside only Australia in Band A, while Band B also includes Austria, Norway, South Korea, Sweden and Taiwan. See Transparency International, <http://government.defenceindex.org/results/overall>, accessed 2 October 2013.

[66] Now located within the System Award Management System, see <https://www.sam.gov/portal/public/SAM/>, accessed 2 October 2013.

[67] David Leigh, Rob Evans and Mark Tran, 'BAE Pays Fines of £285m over Arms Deal Corruption Claims', *Guardian*, 5 February 2010.

[68] Mark Pyman and Tiffany Clarke, 'Raising the Bar: Good Anti-Corruption Practices in Defence Companies', Transparency International, June 2013.

countries have the power to prosecute their nationals for corrupt practices undertaken on other territories.

The central points about these laws are that, while similar in many ways, they have their differences and legal nuances. One difference that is relatively easy to appreciate is that US law allows 'facilitation payments' to encourage an official to do his or her job in an appropriate manner (such as a customs official releasing legitimate goods for import into a country) while British law does not. Thus agents of multinational defence businesses need to know under whose laws they are operating. Second, none of these legal regimes for interdicting bribery and corruption makes any special provision for defence companies. Third, there can be variation in the readiness of national governments actually to prosecute when there is evidence of corrupt practice and to impose significant punishment in the event of conviction.

Conclusion

This chapter has brought together the different aspects by which governments seek to regulate and direct defence companies that are both the suppliers of the key capital assets needed by their armed forces and in receipt of different and varying forms of government support.

As detailed above, the controls imposed by governments have a significant impact on the human resources available to work for defence businesses and on the export markets that they are able to pursue. If it were the case that all eight criteria specified in the European Common Position had to be fully satisfied before a defence export could be allowed, the available international market for Western firms would be restricted indeed. The consequence is that many export-licence decisions hinge on how much weight to place on a specific risk or factor in the light of other considerations. Export marketing thus requires a more or less continuous dialogue between government and business.

The scope of controls over information, employment, exports and behaviour reflect the twin considerations that defence development and manufacturing is far too sensitive and important an area to be left to companies, and yet governments feel obliged to rely primarily on the private sector for these activities because of the efficiency, behaviour and innovation that is associated more with private business than the public sector.

Therefore, whereas for most of the twentieth century the central issue was of national governments trying to direct national firms, the picture today is much richer. National governments want to retain their prerogatives but increasingly they deal with companies that are organised on a transnational basis with some degree of choice about where to locate their development, production and support activities. It is within the bounds of possibility that firms will be drawn to focus more of their activities in countries where the controls are less restrictive, thereby risking reputational damage and legal entanglements as standing hazards of operations within the new business environment.

However, even where the legal situation is clear, the political position may be less clear-cut. Governments do not lightly abandon areas of control. Could Washington legally or politically veto the export of a UK-based subsidiary of a US company if that export comprised only technology developed in the UK? If an American national sits on the UK board of a UK-based US company, could that national be pursued by the US government with regard to the brokering of such an export? The situation in relation to these matters is unclear and may only be clarified through the treatment of problematic cases.

Finally, we would observe that, in each of the countries under scrutiny in this monograph, there are unresolved stresses between the pressures to export defence-related equipment and the desirability and shape of controls. For instance, the US commercial satellite industry may well have suffered particularly from many of its elements being listed under ITAR as opposed to EAR for dual-use items, resulting in lost sales due to the need for an export licence to the country of launch.[69] However, for the UK, US and Germany alike, there is always a complex balance between the risks and benefits of allowing a sale and the commercial consequences of refusing one.

[69] R Zelnio, 'The Effects of Export Control on the Space Industry', *Space Review*, 16 January 2006, <http://www.thespacereview.com/article/533/1>, accessed 2 October 2013.

V. IMPLICATIONS OF THE GOVERNMENT–DEFENCE INDUSTRY RELATIONSHIP

This Whitehall Paper has addressed how government manages its relationship with defence businesses operating in large part from within its territory. The analysis in the previous chapters has conceptualised the role of government through three dimensions: as the customer, sponsor and regulator of such businesses. The United Kingdom, Germany and the United States were used as exemplars of other states due to the applicability of the dynamics between their governments and domestic defence industries. Each country has a significant domestic defence industrial sector, aims to sustain the capability to deploy its armed forces overseas, and has been involved in operations since the end of the Cold War that were difficult to anticipate except shortly before they occurred.

An early finding from this research was that, whilst the challenges facing respective governments are often similar, policy responses in relation to the homeland's defence industrial base remain essentially national. Consequently, how a government responds to these challenges is contingent on a country's political ambition, its culture, history and world view. There is no 'one size fits all' approach, but there are certain factors that must be addressed if a state is to develop an effective relationship with its essential defence businesses.[1] This chapter addresses these factors and the policy implications based on this review.

The analysis below begins by highlighting core conclusions and lessons from the preceding chapters. The case for engagement with ideas generated around the government–defence industrial relationship is then reaffirmed. Thereafter, the discussion outlines the essential *objectives* required across the government–defence industrial relationship within a

[1] See Sean O'Keefe and Gerald I Susman (eds), *The Defense Industry in the Post-Cold War Era: Corporate Strategies and Public Policy Perspectives* (New York, NY: Pergamon, 1998).

major modern state committed to its own defence and security, before progressing to consider the appropriate *characteristics* through which the relationship should prosper.

However, before analysing these conclusions, a simple, but essential, point must be highlighted. The issue of government–defence industry relations is critically important to all citizens and should not, indeed cannot, be ignored.[2] The complexities of the twenty-first-century geopolitical system and the changing national-security environment drive this imperative. While it is clear that military forces cannot absolutely ensure the protection of populations against every type of malicious threat, as terrorist attacks around the world demonstrate, they are a key element of national security including the counter-terrorist effort, and their performance, agility and resilience is much determined by their supply base: that is, the defence industry.

Some Core Lessons

Despite the often tired free-market narrative surrounding the UK government's acquisition of defence equipment and services 'commercially, off-the-shelf', through a series of 'open competitions' embracing a 'global supply chain',[3] it is clear that governments across the world – as a function – still sponsor, or champion, their own national defence industrial base as best they can. This is quite apparent from even superficial scrutiny of countries as different as Brazil, China, Turkey, India and the UAE. Within this broader point, states develop their own systems and husbandry processes for the defence businesses operating within their borders. To this extent, such a practice is a national imperative varying in substance and purpose from polity to polity, but clearly identifiable as a function of statehood. The existence and safeguarding of defence businesses, the promotion of their commercial interests, and the provision of skills, competencies and other scarce and precious elements of a defence value-chain that generates specialist defence goods and services are inescapably the business of

---

[2] For an extensive discussion of this point see Aerospace Industries Association, 'The Unseen Cost: Industrial base Consequences of Defense Strategy Choices', July 2009.

[3] See, in particular, MoD, *National Security Through Technology: Technology, Equipment, and Support for UK Defence and Security*, Cm 8278 (London: The Stationery Office, February 2012); see also the analysis of this document by the RUSI Acquisition Focus Group in *RUSI Defence Systems* (Vol. 14, No. 3, Spring 2012), pp.14–16.

government. This monograph demonstrates that it could hardly be otherwise, irrespective of the political fashions of the moment.

Through the defence industrial triptych of customer, sponsor and regulator, a mutually reinforcing and interdependent relationship can be discerned among government, a country's military component and its defence industrial base, formed of large and small businesses operating from its lands.

In some ways, where the ultimate headquarters of that business happens to be registered or located scarcely seems to matter anymore given that the location of capital assets and the knowledge and skills of a local workforce are of central importance. Certainly, multinational defence companies are aware of the potential advantages of their structure, with firms able to adopt the optimum national flag under which to operate in international defence exhibitions and markets. Thales and BAE Systems are among those firms which have sought with some success to present themselves as national entities in the several countries which they designate as 'home markets'.

Nonetheless, for governments and companies this multinationality presents some current and potential issues, especially in terms of sponsorship and regulation in Europe. Is the French government content for research work that it funds with Thales to be transferred to the UK? How much is the efficiency and profitability of a firm such as Thales or MBDA impeded by national restrictions on access to information for individuals or the transfer of technology? Should the Defence and Security Organisation (DSO), which promotes the export of UK goods and services, support the sale of BAE Systems armoured vehicles that would be made predominantly in Sweden? What is the cost entailed in a firm's requirement for export licences in order to move a component or sub-system from one site of the business in one country to another company site abroad?

These issues have particular salience within Europe, where governments are interested in building defence businesses that are large enough and sufficiently well-resourced to be able to compete (and collaborate) with their US equivalents. While the questions are relevant to European businesses operating in the US, the clarity of issues is enhanced by the very strict controls that the US places on the export of information and technology. Companies investing in the US cannot expect to be allowed to transfer US-origin technology to Europe (unless it is in connection with a major export sale). Even then, they will not be able to re-use that technology in other areas without US government permission.

However, the regulation of defence exports raises questions for all three countries considered in this monograph. Could Lockheed Martin UK, which has ownership of substantial amounts of UK-origin technology, export that technology to a market of which Washington did not approve? Could BAE Systems, Inc (the US company) sell products to a country of concern to the British government? There are no clear rules in these areas and it may take decades, and the emergence of some challenging cases, for clear guidelines to emerge. In the meantime, there is, of course, the possibility that companies will opt to invest in research, development and production in countries where the export regulation is seen as being the lightest.

Established practice with European collaborative projects may offer some hint as to what will occur. Although the details of the Memoranda of Understanding establishing collaborative development and production projects are normally classified, it is the authors' understanding that they usually contain a clause that would enable any partner to veto an export. However, there is also awareness within governments that anything except the rarest exercise of such a veto would much reduce a country's appeal as a collaborative partner for the future. Hence it has become implicit practice that if one state deems an export to be reasonable, the other partners will not query that choice. An illustration would be the sale of Tornado to Saudi Arabia under the original Al-Yamamah programme in the mid-1980s. Germany then had a more restrictive arms-export policy than the UK and would not have sold such an aircraft to Saudi Arabia had it been a national product. Indeed, it decided around that time not to sell Leopard tanks to Saudi Arabia. It was, however, content to allow the UK to lead on the sale. Likewise, the Franco-German HOT and MILAN missiles were sold to many states, with France very much in the lead sales role.

In brief, national governments must not only draw their sponsorship, regulation and customer roles into an integrated whole, but they must do so while dealing with businesses that are no longer national in ownership or strategic direction.

Understanding these relationships and grasping their implications is now the business of the politician, industrialist, warrior and citizen. Such an understanding should frame policies and strategies – military, economic and industrial – as they navigate the choppy waters of this century. Offered below are some suggestions for the values that should be promoted and objectives that should be sought in the process.

The Case for Engagement

Why does government's relationship with its on-shore defence industrial base merit special scrutiny in the twenty-first century? In the US, the UK and Germany, the national defence and security effort has reached what Jacques Gansler, a former US secretary of defense for acquisition, technology and logistics, describes as the 'tipping point' which should prompt profound policy soul-searching and a case for significant change.[4] The full range of national-security considerations and matching capabilities are increasingly unaffordable in an era when the nation must spend increasing proportions of its income on health and social security, infrastructure renewal and the ongoing repayment of structural debts. Consequently, defence programmes need to be delivered as effectively and efficiently as possible if traditional defence stances are to endure. Analysts and politicians also point to growing asymmetrical threats and new tools of conflict, such as cyber-attacks, which could render the state impotent in the face of a strike.[5] Together, the fiscal realities faced by states and the range of new security challenges they must confront mean that things have to be done differently and the case for change becomes powerful. Effective use of indigenous defence industries seems a sensible and important component of an appropriate policy response, providing it can be done systematically and in an affordable manner.[6] The case for engagement, therefore, is made by the perfect storm of security challenges combined with the fiscal issues faced by the modern state (and budgetary challenges for individual ministries). Within this context, what does the state require from an appropriate defence industrial structure if the citizen is to stay safe and secure?

Defence Industrial Objectives for the Twenty-First Century

In future, the relationship between government and industry must ensure that:

- Defence industries are *responsive to an ever-changing broad sweep* of national-security threats, risks and issues
- Defence industries become *highly agile*, with integration between civil and military technologies, and their use, becoming the norm
- The *unit costs* of defence capabilities should fall as cost reductions are brought about from greater through-life efficiencies.

---

[4] Jacques S Gansler, *Democracy's Arsenal: Creating a Twenty-First-Century Defence Industry* (Boston, MA: MIT Press, 2011), p. 358.

[5] See, for instance, Bernard Jenkin and George Grant, 'The Tipping Point: British National Strategy and the UK's Future World Role', Henry Jackson Society, July 2011; Alvin Toffler and Heidi Toffler, *War and Anti-War: Survival at the Dawn of the 21st Century* (New York, NY: Little Brown, 1993); Pascal Bruckner and Steven Rendall, *The Tyranny of Guilt: An Essay on Western Masochism* (Princeton, NJ: Princeton University Press, 2006).

[6] See George Friedman, *The Next Hundred Years: A Forecast for the Twenty-First Century* (London: Allison & Busby Ltd, 2009).

Each of these issues is now considered in turn. First, the support to government that commercial defence businesses provide will be pursued, in future, across a broad sweep of defence and security threats, issues, risks and responding requirements.[7] This support will range from the generation of traditional capital items, such as nuclear-powered submarines, through to problems such as water security and the interdiction of cyber-threats against domestic economic and social targets. The defence industrial and security base will, therefore, continue to morph, expand and contract in response to a dynamic and complex security situation. In complexity, small changes can lead to large changes in outcome. Consequently, industry will have to satisfy a broad spectrum of national-security needs beyond its traditional nineteenth- and twentieth-century roles of manufacturing items for conventional war. Governments will need to plan the effective utilisation of national resources to maximise the widest possible portfolio of expertise, now residing within the private sector, to meet this spectrum of requirements. This is something that will require *both* deliberate centralised planning and the dynamism and discipline of the free market, as well as the refinement and articulation of inherently government functions.[8]

Second, the defence industrial structure will have to be highly agile to meet the uncertainties and risks of the future national-security environment. A government's role will be the generation of policies and practices to engrain and ensure this agility.[9] Competitors and adversaries, both state and non-state, will be able to acquire technology rapidly and generate new threats by exploiting the global technology marketplace. The knowledge contained within the national defence industrial base must reside within a structure that allows an effective and continual responsiveness to rapidly emerging technological threats. Part of this challenge will involve the removal of barriers to integration that exist between the civil and military spheres, whilst recognising that governments, and corporations, will always want to protect (or restrict access to) certain sensitive technologies. Nonetheless, open architectures and rapid integration should be the default position of this century's national defence industrial structure.[10]

---

[7] Burkhard Schmitt (ed.), 'Between Cooperation and Competition: The Transatlantic Defence Market', Chaillot Paper No. 44, ISS-WEU, 2001.

[8] See Stephen G Brooks, G John Ikenberry and William C Wohlforth, 'Lean Forward: In Defense of American Engagement', *Foreign Affairs* (January/February 2013); Trevor Taylor and John Louth, 'What the Government Must Do in Defence Procurement', RUSI Briefing Paper, October 2013.

[9] This point was made by Pierre Chao in a speech he delivered to the US Industrial College of the Armed Forces on 2 June 2005, entitled 'The Future of the US Defense Industrial Base: National Security Implications of a Globalized World'.

[10] See Ethan B Kapstein, *The Political Economy of National Security: A Global Perspective* (Columbia, SC: University of South Carolina Press, 1991).

Third, for defence and security capabilities to be useful they must be sustainable, which means, in turn, that they must also be affordable.[11] Efficiency and affordability must be built into the programme solution from the moment a requirement is identified. As Gansler suggests:[12]

> To afford the equipment that will be required for potential future security scenarios (both domestic and worldwide) the industrial structure needs to reduce the unit costs of the equipment dramatically. The current costs of single ships or planes, for instance, are prohibitive. To get the quantities required in the future, lower costs, through both product and process designs, must be a firm military requirement for all future weapons systems and systems of systems.

This notion of affordability should also focus on the unit costs of the capability rather than the national budget for the equipment development and delivery alone.

This grouping of affordability, agility and breadth will be the defining value-set of this century's successful national defence industrial asset base. Government's role should be to use the levers associated with being a customer, sponsor and regulator of its defence and security industry to enable this end state. Achieving these desired characteristics could involve a transformation of the existing defence industrial structure within countries, but the necessary change to align potential responses to potential threats is the responsibility of government. The practices of government as customer, sponsor and regulator therefore must reflect, in future, this critical national-security responsibility.

A Responsive Government–Defence Industrial Relationship

As a consequence of these objectives, what characteristics should an effective and responsive government–defence industrial relationship possess? The principal, and perhaps most critical, factor is government's clarity around what it actually requires from defence businesses, with this explicability and insight positively informed by industry. Government needs to know *what capabilities* it wishes to generate or refresh; the *contracting mechanisms* for securing the required elements from the industrial base; *who* will co-ordinate, source and integrate the ingredients that make up these identified capabilities; and *how* systems can be acquired that maximise performance, minimise unit costs and overall total programme costs, and can be delivered within the least amount of time.

A state defence and security acquisition system endowed with these qualities is central to a successful government–defence industrial

[11] See Malcolm Chalmers, 'Mid-Term Blues? Defence and the 2013 Spending Review', RUSI Briefing Paper, 2013.
[12] Gansler, *Democracy's Arsenal*, p. 346.

relationship. Indeed, the reason for the recent plethora of reform activity in the defence acquisition processes in the UK, Germany and the US is the perennial search for these qualities. Yet going forward, some degree of stability and certainty may be necessary across the process if effective, efficient and affordable defence and security capabilities are to emerge. This stability might revolve around the application of the following characteristics.

Characteristic 1: Net-Centric Capabilities
As demonstrated throughout this Whitehall Paper, the relationship between government and the defence industry has matured over the years through the very specific paradigms of government as customer, sponsor and regulator. This has necessitated, historically, a focus on platforms and particular pieces of equipment, or bounded packages of service provision. The twenty-first-century security environment suggests that one must now think in terms of complex and interdependent systems, systems of systems, or net-centric capabilities, whilst also ensuring that the focus on low total and unit costs is cognisant of this 'systems-view' of capabilities. This is both conceptually and programmatically challenging; in the modern order of battle, equipment packages or 'at border' security platforms merely become nodes of a wider integrated defence and security capability. This necessitates a change in mindset around such things as requirements setting, budgeting and capability planning, shifting the focus from projects, and even from programmes, to portfolios. Government and industry should reform their relationship around this necessity, conversing through a language of portfolios and systems rather than platforms and items of equipment. It would be a profound change in emphasis.

Characteristic 2: Oversight
The relationship between government and commercial businesses involved in defence and security is complex and governed by certain bespoke norms, expectations and practices. It is an important relationship through which the state secures its borders and projects capabilities onto the lands of others. This complexity and importance suggests that the relationship should not be left to 'free-wheel' but should be overseen by a body explicitly charged with its protection. It is the authors' proposal that the best solution in this regard is located in the legislative branch.

Of course, defence and security committees of parliamentarians already exist in Washington, Berlin, London and elsewhere; there is the UK Defence Select Committee, for instance, and the US Senate and House Armed Services Committees. None, though, overtly see their role as including oversight and stewardship of the relationship between government and the capabilities residing within the defence industrial base, or see this as a shortfall. The nearest elements are the Readiness Subcommittee of the House Armed Services Committee and the Readiness and Management Support Subcommittee of the Senate Committee on the Armed Services in the US. Parliaments, through their committee structures, should continually conduct an audit or stock-take of the critical national defence and security capabilities dwelling within the commercial sector and periodically measure the state of their health and readiness.[13] This analysis should be a standing committee order of business, so that the efficacy of the national government–industrial relationship across the defence and security sectors is subject to continual, constitutional scrutiny and oversight.

Characteristic 3: Research and Development

Chapter III discussed how the United States' strategy for its own national security has been to maintain technological superiority over its political foes. Likewise, British policy since 2012 is to protect research, applied research and development by ring-fencing 1.2 per cent of the annual defence budget for the purpose of spending on science and technology. Research is perceived as a critical function of defence and national security, and is considered a series of activities undertaken within government departments, across universities and throughout the defence industrial base. Yet much of this research is private and takes place within stovepipes, with research teams often ignorant of their colleagues' efforts.

A significant advance would be for both government and commercial entities to accept that fundamental research should be publishable and available to all through open-access protocols. Pure research should be transferable without boundaries or protective caveats, and only 'monetised' and treated as intellectual property once it enters the development and manufacturing cycles of a specific programme.[14] Such an approach would enable research collaborations between businesses and across public and private sectors, driving a multi-use approach between defence and civil sectors.

[13] Whilst shadowing the activities of government is a massive task, oversight in this manner contributes towards the health of a system that has government as a constituent part.

[14] The desirability of this characteristic was recognised by President Reagan in the 1980s when he issued the US National Security Decision Directive 189. This stated that research should be 'open-booked' and not protected. Unfortunately, despite the directive, this stance was never properly operationalised.

Characteristic 4: A Skilled Acquisition Workforce

The relationship between government and the defence industry – whether through the lens of customer, sponsor or regulator – in many ways is 'operationalised' through the activities of the defence acquisition workforce. Consequently, the skills and competencies contained within this function are subject to significant debate and review, which reveal potential differences in approaches between states. There is, no doubt, consensus on the notion that a core characteristic of a functioning relationship between government and industry is the generation and sustainment of a skilled and effective defence acquisition workforce, of sufficient critical mass, positioned as the virtual hinge between the government component and the industrial and service elements of national security. Within the US and Germany, there is a policy preference to source, develop and mature these skills within the public sector as 'inherently governmental' functions. By contrast in the UK, a business case is being prepared by Bernard Gray, chief of defence materiel, for a decision in 2014 to outsource defence acquisition to the private sector. Consequently, whilst states are unified in the recognition of the importance of the defence acquisition function, there is a potentially significant divergence in response, at least on the part of the British. For most advanced industrial states, functions such as programme and project management, commercial negotiations and contracting, financial control and budgeting, risk and opportunity management, logistics and assurance are all necessary extensions of government, especially in terms of defence and national security. Within the UK, the government suggests that as these skills are best found in the private sector, they should be sourced from this sector through the de facto privatisation of defence acquisition.[15]

It is a major cleavage in policy response as suggested by US Undersecretary of Defense for Acquisition, Technology and Logistics Frank Kendall:[16]

> My counterpart in the UK, Bernard Gray, has ... come to the conclusion that he does not have the expertise in government ... to allow him to manage programmes from within government. He's going to go out and hire somebody, essentially, to do that for him.

[15] See Ministry of Defence, *Better Defence Acquisition: Improving How We Procure and Support Defence Equipment*, Cm 8626 (London: The Stationery Office, June 2013); RUSI Acquisition Focus Group, 'The Defence Materiel Strategy and the GOCO Proposal for Abbey Wood', RUSI Briefing Paper, July 2012.

[16] 'Better Buying Power Initiative 2.0', speech by Frank Kendall, US undersecretary of defense for acquisition, technology and logistics, at the Center for Strategic and International Studies, Washington, DC, 14 November 2012.

> I don't want to do that. I think the government needs to have the in-house expertise to do the business management [and] the technical management... The single most important determiner of outcomes is the quality of leadership, period.

Whilst the approaches to the subject vary and are plainly contested, an effective government–defence industry relationship rests on the effective performance of the acquisition process and the ability of government to be a smart customer, an insightful sponsor and appropriate regulator of industrial efforts and ambitions. In this regard, it is perhaps the defining characteristic of a successful relationship.

Characteristic 5: The Deployed Defence Contractor

In the twenty-first century, defence businesses do not just provide equipment and services for utilisation by the state's military forces but also contribute directly to capabilities being deployed on the front line. Based on input costs, contractor support to operations (CSO) comprises at least 40 per cent of the military effort on deployed operations such as Iraq and Afghanistan. Indeed, for the UK, £2.6 billion was spent in 2010 on personnel from industries employed in support roles with the military on operations.[17] Contractors employed by the United States comprised half of its total deployed force to Iraq whilst, also for 2010, 239,451 contractors (including locally employed Afghan nationals) were employed by the US in Afghanistan. This represented 75 per cent of the in-theatre deployed force.[18] While many of these contractors are engaged in providing services for troops within bases (such as catering and laundry) and the protection of logistics deliveries, some provide equipment support and in a few cases help with the operation of (surveillance) equipment.

By most indicators, business is now a fully-fledged element of the military component without which many sustained deployed operations are unthinkable, let alone deliverable. This consideration will characterise the government–defence industrial relationship of the future and should form part of the policy lexicon going forward. Elements of war-fighting, peace-making and peacekeeping are now subject to state outsourcing, which means that ethical considerations and governance arrangements need proper attention. Yet, at present, and despite the experiences of recent operations, remarkably this is not high on the agenda of political parties across the US, the UK, Germany and elsewhere.

---

[17] See Henrik Heidenkamp, 'Sustaining the UK's Defence Effort: Contractor Support to Operations Market Dynamics', *Whitehall Report 1–12* (April 2012).

[18] See Richard Fontaine and John Nagl, 'Contracting in Conflicts: The Path to Reform', Center for a New American Security, June 2010; Heidenkamp, 'Sustaining the UK's Defence Effort'; Christopher Kinsey, *Private Contractors and the Reconstruction of Iraq* (London: Routledge, 2009).

Of course, it is recognised that in advanced liberal democracies the development – or, more properly, the endorsement – of policies is the preserve of political parties seeking to govern a state. The core point is that, when generating those policies, political actors must take account of these essential values and characteristics when engaging with defence businesses, whether as their customer, sponsor or regulator. Failing in this task will leave countries and their citizens potentially exposed to the salient threats, risks and hazards of an uncertain century.

Conclusion

Moving from the current practices of a government–defence industry relationship driven by roles of 'customership', sponsorship and regulation to a responsive relationship framed for the security challenges of the twenty-first century is not without challenges. However, recognising the need for a change in direction and knowing where one happens to be positioned today are the two key precursors to any successful transformation.[19] To this end, what of the questions posed at the end of the Introduction:

- Do national policies across the three areas of 'customership', sponsorship and regulation enjoy coherence, or can contradictory elements be discerned?
- Do government constraints seriously hinder the ability of defence enterprises to operate effectively and efficiently?
- Do governments' stances towards the defence industry facilitate or bring friction to collaborative and co-operative projects?
- Do governments' stances towards the defence industry encourage it to invest in some countries rather than others?
- What directions of change can be discerned in governments' stances towards the defence industry?
- Do governments aspire to control (and/or sponsor) the overseas investments of defence companies based on their territory?
- Does government policy and behaviour recognise the implications of and questions raised by cross-border investments in the defence industrial sector?

Through the very specific perspectives of the United Kingdom, the United States and Germany, the analysis above establishes that government engages with defence businesses through multiple roles captured under the banners of customer, sponsor and regulator. It is clear, though,

[19] For a discussion on the dynamics of transformation and the practices of change management see John P Kotter, 'Leading Change: Why Transformation Efforts Fail', *Harvard Business Review on Change* (January 2007).

that it does not do this in a coherent way; for example, no governmental policies or political party manifestos appear to have been derived under these three headings or anything similar. As a consequence, and perhaps inevitably, inconsistencies and contradictions can be perceived.

In the United Kingdom, the government wants to be a customer of defence industries through a process of open competition as the preferred contracting mechanism. Yet it also sponsors, to the point of exclusivity, industry sub-sectors such as submarine design, manufacture and maintenance and on-shore small-arms ammunition manufacturing so that, despite a preference for competition, some 40 per cent of the UK's defence acquisition portfolio by value and around 65 per cent by number of contracts involves non-competitive procurement.[20]

In Germany, despite a strong public narrative promoting overt government regulation of defence and security capabilities, conventional market dynamics can be discerned. The proposed BAE Systems-EADS merger of 2012 clearly took the German government by surprise despite the operation of a supposedly engaged regulatory regime. Without an open and clear recognition by the political parties of advanced states that government undertakes, and has always performed, these three key roles, it is inevitable that a lack of coherence will be the mark of government efforts to be an informed and knowledgeable customer, an appropriate sponsor seeking responses from industry to counter amorphous new century threats, or a sophisticated regulator of lethal capabilities. It would be more shocking to conclude that there was an element of alignment between the three roles; which, the authors suspect, will not be possible for many years to come.

That is not to say that government impedes the activities and ambitions of defence businesses. As far as overseas investment is concerned, that appears to be largely a matter for companies to decide rather than their home governments providing direction: governments worry most about what defence businesses are being sold inside their territory. However, it is clear that, in the Gulf in particular, the British government and UK defence companies are working closely to develop a larger significant UK role in the area, to promote UK defence sales and to take seriously ideas about the development of indigenous Gulf defence businesses.

On the business side of the equation, no indicators were found that industrialists wish to operate in a perfect market beyond any government constraint or engagement. Rather, businesses operating in this market recognise the unique governmental framework that enwraps it as a

20 DASA, 'UK Defence Statistics Compendium', 2012, <http://www.dasa.mod.uk/publications/UK-defence-statistics-compendium/2012/chapter-1-finance/chapter-1-finance.pdf>, accessed 8 October 2013.

necessary market condition of operation. Bureaucracy is seldom viewed positively and there are perennial demands to reduce 'red tape', but industrialists welcome government engagement in this market through the three roles identified. If anything, such engagement reduces the risks to the business and its investors by mapping and confirming the topography of the market. As one senior industrialist commented:[21]

> The government buys our products, initiates our order pipeline, sponsors the refreshment of skills for the business, and acts as our sales champion for exports. Government is the enabling condition of the market rather than a constraint [on its] operation ... Without government's guiding hand and co-ordination there would be no major collaborative or international projects like the F-35 programme, and little innovation without government's sponsorship of pure research.

Government, therefore, does not hinder innovation or collaboration: instead, it is an important element within a complex defence ecosystem and a component that appears to generate and enable co-operation between businesses and across state boundaries.

In contrast, arguments around internal investment by defence businesses are more nuanced. Investors expect businesses to generate a return on their investments, with the focus inevitably falling on sales and turnover, margin and capital growth. Countries that actively sponsor and purchase defence capabilities can reasonably expect defence businesses to inject their investment dollars, pounds and euros into their national industries. Research activities and production processes are frequently clustered so that specific economic benefits are felt both locally and nationally. As a consequence, the promotion and protection of defence jobs become part of the narrative of the relationship between government and defence industry, a narrative that can seriously energise the media. However, defence industries also invest, through their marketing and sales functions, in countries they perceive to be important to tomorrow's growth. A defence company's sales and opportunities pipeline embraces both current relationships with governments and those that could emerge in the future, inevitably backed by the diplomatic efforts of the host state. Yet the benefits of new orders, future collaborative projects and exports are of course felt through tax yields in the country within which the business is registered, making overseas investment potentially beneficial to multiple countries. Government cannot control these investments but can, and does, champion their occurrence, both to influence the behaviour of the benefitting state and for the associated economic benefits at home. It is an economic argument as much as a diplomatic or capability discussion.

---

21 Author interview with senior defence industrialists, London, 19 March 2013.

Throughout this monograph the centrality of national defence industrial and service capabilities to defence and national security has been continually underlined. This is manifested through a complicated and contested relationship that exists between government and the businesses themselves. As extant capabilities continue to reside in the industrial base and as new capabilities such as unmanned air, land, sea and submarine systems are operated from within industry (not to mention new cyber-offensive and defensive capabilities), the private sector will not just be the repository of much of the armoury of advanced states, but it will also be the operational partner for many force elements. This means that the relationship between government and this critical component of defence capability needs to be better understood and subjected to significantly more clear-eyed policy-making. The age of the private army has yet to dawn but a new reality now exists, with the private sector fielding many elements of defence capability. This Whitehall Paper suggests that governments and their peoples are not, as yet, cognisant of the responsibilities that this change in perspective generates, let alone united in a response.

About Whitehall Papers

The *Whitehall Paper* series provides in-depth studies of specific developments, issues or themes in the field of national and international defence and security. Published occasionally throughout the year, *Whitehall Papers* reflect the highest standards of original research and analysis, and are invaluable background material for specialists and policy-makers alike.

About RUSI

The Royal United Services Institute (RUSI) is an independent think tank engaged in cutting-edge defence and security research. A unique institution, founded in 1831 by the Duke of Wellington, RUSI embodies nearly two centuries of forward thinking, free discussion and careful reflection on defence and security matters.

RUSI consistently brings to the fore vital policy issues to both domestic and global audiences, enhancing its growing reputation as a 'thought-leader institute', winning the Prospect Magazine Think Tank of the Year Award 2008 and Foreign Policy Think Tank of the Year Award 2009 and 2011. RUSI is a British institution, but operates with an international perspective. Satellite offices in Doha and Washington, DC reinforce its global reach. It has amassed over the years an outstanding reputation for quality and objectivity. Its heritage and location at the heart of Whitehall, together with a range of contacts both inside and outside government, give RUSI a unique insight and authority.